Greenland and the International Politics of a Changing Arctic

Greenland and the International Politics of a Changing Arctic examines the international politics of semi-independent Greenland in a changing and increasingly globalised Arctic. Without sovereign statehood, but with increased geopolitical importance, independent foreign policy ambitions, and a solidified self-image as a trailblazer for Arctic indigenous peoples' rights, Greenland is making its mark on the Arctic and is in turn affected – and empowered – by Arctic developments.

The chapters in this collection analyse how a distinct Greenlandic foreign policy identity shapes political ends and means, how relations to its parent state of Denmark is both a burden and a resource, and how Greenlandic actors use and influence regional institutional settings as well as foreign states and commercial actors to produce an increasingly independent – if not sovereign – entity with aims and ambitions for regional change in the Arctic.

This is the first comprehensive and interdisciplinary examination of Greenland's international relations and how they are connected to wider Arctic politics. It will be essential reading for students and scholars interested in Arctic governance and security, international relations, sovereignty, geopolitics, paradiplomacy, indigenous affairs and anyone concerned with the political future of the Arctic.

Kristian Søby Kristensen is Deputy Director and Senior Researcher at the Center for Military Studies, Department of Political Science, University of Copenhagen. With a background in international relations, his academic interests are war and strategy, Arctic politics, Danish and European security and defence policy as well as issues of public safety and security.

Jon Rahbek-Clemmensen is Assistant Professor at the Department of Political Science and Public Management at the University of Southern Denmark, where he is affiliated to the Center for War Studies. He holds a PhD in international relations (LSE) and has been affiliated to Columbia University and CSIS. His research interests include European security, Arctic politics, Danish foreign and security policy, and civil–military relations.

Routledge Research in Polar Regions

Series Edited by Timothy Heleniak

The Routledge series Research in Polar Regions seeks to include research and policy debates about trends and events taking place in two important world regions: the Arctic and Antarctic. Previously neglected periphery regions, with climate change, resource development and shifting geopolitics, these regions are becoming increasingly crucial to happenings outside these regions. At the same time, the economies, societies and natural environments of the Arctic are undergoing rapid change. This series seeks to draw upon fieldwork, satellite observations, archival studies and other research methods which inform about crucial developments in the polar regions. It is interdisciplinary, drawing on the work from the social sciences and humanities, bringing together cutting-edge research in the polar regions with the policy implications.

For a full list of titles in this series, please visit www.routledge.com/Routledge-Research-in-Polar-Regions/book-series/RRPS

Recently published titles:

New Mobilities and Social Changes in Russia's Arctic Regions
Edited by Marlene Laruelle

Climate, Society and Subsurface Politics in Greenland
Under the Great Ice
Mark Nuttall

Greenland and the International Politics of a Changing Arctic
Postcolonial Paradiplomacy between High and Low Politics
Edited by Kristian Søby Kristensen and Jon Rahbek-Clemmensen

Greenland and the International Politics of a Changing Arctic

Postcolonial Paradiplomacy between High and Low Politics

Edited by Kristian Søby Kristensen and Jon Rahbek-Clemmensen

Routledge
Taylor & Francis Group

LONDON AND NEW YORK

First published 2018 by Routledge

2 Park Square, Milton Park, Abingdon, Oxfordshire OX14 4RN
52 Vanderbilt Avenue, New York, NY 10017

Routledge is an imprint of the Taylor & Francis Group, an informa business

First issued in paperback 2019

British Library Cataloguing-in-Publication Data
A catalogue record for this book is available from the British Library

Library of Congress Cataloging-in-Publication Data
A catalog record for this book has been requested

ISBN: 978-1-138-06109-5 (hbk)
ISBN: 978-0-367-36234-8 (pbk)

Typeset in Times New Roman
by HWA Text and Data Management, London

Contents

Figures

Contributors

Klaus Dodds is Professor of Geopolitics at Royal Holloway, University of London, UK and his research interests include geopolitics, security and the governance of the polar regions. He is co-author with Mark Nuttall of *The Scramble for the Poles* (Polity 2016) and co-editor of *Handbook on the Politics of Antarctica* (Edward Elgar 2017).

Kevin Foley is a PhD student in the Department of Government at Cornell University, USA. His research interests include the role of ideas and political rhetoric in the foreign and security policy of advanced democracies and the politics of China and southeast Asia.

Ulrik Pram Gad is Associate Professor at Aalborg University, Denmark. He directs a project on the *Politics of Sustainability in the Arctic*. Recent publications include *National Identity Politics and Postcolonial Sovereignty Games* (Museum Tusculanums Forlag 2016) and a special issue of *Security Dialogue* on 'The politics of securitization' (with K.L. Petersen).

Hannes Gerhardt is Associate Professor of Geography at the University of West Georgia, USA. His research focuses on political geographic imaginaries, particularly as they pertain to the spatial ordering of political power, capitalism, and alternatives to capitalism.

Jens Heinrich holds a PhD in history from the University of Greenland (Ilisimatusarfik). The former vice president of the Greenland Reconciliation Commission, he currently works as a political consultant for the Inuit Ataqatigiit party in the Danish Parliament. His academic work has focused on the historical development of Greenland and the relationship between Greenland and Denmark.

Marc Jacobsen is a PhD student at the Department of Political Science, University of Copenhagen. Denmark. His main research interest is how Denmark and Greenland use the Arctic to discursively position themselves internationally.

Inuuteq Holm Olsen is First Minister Plenipotentiary and Head of Representation for Greenland at the Danish Embassy in Washington, DC. An experienced diplomat, he has been Greenlandic Deputy Minister for Foreign Affairs and Senior Advisor in the Danish Ministry of Foreign Affairs. He holds an M.A. in international affairs from George Washington University, USA.

Kristian Søby Kristensen is Deputy Director and Senior Researcher at the Center for Military Studies, Department of Political Science, University of Copenhagen, Denmark. His research interests include NATO, Western strategy and security, Danish defence, security and foreign policy, resilience, Arctic governance and security.

Mark Nuttall is Professor and Henry Marshall Tory Chair of Anthropology at the University of Alberta, Canada. He also holds a visiting position as Professor of Climate and Society at the University of Greenland and Greenland Institute of Natural Resources. Much of his current research focuses on climate change, extractive industries and human–environment relations in Greenland.

Mikkel Runge Olesen is a senior researcher in the Foreign Policy Unit at the Danish Institute for International Studies (DIIS). His research interests include Danish foreign and security policy, transatlantic relations, and Arctic politics.

Jon Rahbek-Clemmensen is Assistant Professor at the Department of Political Science and Public Management at the University of Southern Denmark, where he is also affiliated to the Center for War Studies. His research interests include European security, Arctic politics, Danish foreign and security policy, and civil-military relations.

Jessica M. Shadian is Nansen Professor at the University of Akureyri, Iceland and Senior Fellow at the Bill Graham Centre for Contemporary International History, Toronto, Canada. Her research focuses on Arctic resource governance and law, Inuit governance, the role of the EU in Arctic affairs, and the politics of Arctic science.

Camilla T. N. Sørensen is Assistant Professor at the Institute for Strategy at the Royal Danish Defence College, Denmark. Her research interests include Chinese foreign and security policy, East Asian security, Arctic politics, and Danish foreign and security policy.

Acknowledgements

The idea for this volume grew out of our engagements with an increasing international – and indeed global – audience interested in a changing Arctic. As Danish academics working on Arctic politics, we have each experienced being asked to present a/the 'Danish' view on Arctic politics. These presentations frequently wound up in complex, but inevitably all-too-hasty, explanations of precisely who runs Greenland's foreign relations, how they are (not) authorised to do so within the constitutional framework of the Danish Realm, the complex identity dynamics underpinning Greenland's relationship with the outside world, and why they are central to understand changing Arctic international politics.

Hopefully, this book cuts through this complexity and conveys a clearly understandable overview of how Greenland operates in the international sphere and how the international sphere concurrently shapes Greenland. With this book we bring together scholars interested in Arctic studies, international relations, postcolonialism, identity politics, and geopolitics to help us show an international audience how to understand and analyse the importance of Greenland's international relations in a changing Arctic. Instead of trying to cover the changing politics of the entire Arctic, our aim is to thoroughly explore Greenland's international relations to see what we can learn of Arctic politics through that investigation.

Collected volumes are by definition collaborative endeavours. And a number of people and institutions deserve to be acknowledged for their help and support. Most importantly, our contributors have all showed great enthusiasm and diligently responded to our editorial queries. An important milestone for the project was a workshop held November 2015 in Copenhagen, and we want to thank the participants there for their valuable input and for fruitful discussions on the Arctic and Greenland. The same goes for the people present at an Arctic politics seminars organised by Ulrik Pram Gad at the Department of Political Science, University of Copenhagen, where we presented a draft of our introductory chapter.

From the moment we approached series editor Timothy Heleniak of the Routledge Research in Polar Region series, we have been very well supported and encouraged by the Routledge team. Thanks to Faye Leerink and Priscilla Corbett for all your help with the publication process and for quick replies to not always easy questions. We are equally grateful to the reviewers for their very helpful comments and suggestions both to the volume as a whole and to the individual chapters.

We finally wish to acknowledge that this research project is part of the Center for Military Studies' research-based services for the Danish Ministry of Defence, and that the project could not have been possible without financial support from the Carlsberg Foundation.

Kristian Søby Kristensen
Jon Rahbek-Clemmensen
Copenhagen and Odense

March 2017

Introduction: Greenland and the international politics of a changing Arctic

Postcolonial paradiplomacy between high and low politics

Kristian Søby Kristensen and
Jon Rahbek-Clemmensen

On 28 May 2008, high-level representatives from the five Arctic coastal states (Canada, Denmark/Greenland, Norway, Russia, and the United States) met in Ilulissat, Greenland for what would become a pivotal meeting in the political history of the region. It was a time of turmoil and uncertainty in the High North. The Russian flag planting on the North Pole the previous year had sparked tensions between the High North capitals, and experts and commentators warned that an Arctic great game for resources and territory could be under way (Chivers 2007; Borgerson 2008). The regional capitals also worried that the frequent calls for an Arctic treaty, following the Antarctic model, would internationalize the region and move power away from the coastal states (Hertell 2008; Koivurova 2008). In Ilulissat, the states agreed to a regional framework that neutralized many of their fears and anxieties, as the Ilulissat Declaration essentially placed matters pertaining to the Arctic Ocean in the hands of the five coastal states, and underlined that regional institutions, such as the Arctic Council, would remain crucial forums for facilitating regional inter-state cooperation (The Ilulissat Declaration 2008).

The standard story of the Ilulissat Declaration is essentially about state power. It was states that had the influence and the ambition to take the lead in instituting an order that benefitted themselves at the expense of other actors. It is often forgotten that the ministerial meeting was the result of a joint initiative and invitation by the Danish *and* the Greenlandic Government. Greenland, a self-governing nation within the Kingdom of Denmark with a realistic path to independence, could see itself in an order that emphasized states. Greenlandic paradiplomacy – the diplomacy of sub-state regional entities vis-à-vis foreign actors – influenced and gave credibility to the Danish initiatives and thus came to play a crucial role in a vital ordering moment in Arctic politics. As an Inuit subnational polity, Greenland lent legitimacy to the Ilulissat initiative, by showing that it did not simply originate in capitals far south of the Arctic Circle, but that at least one indigenous people of the High North stood behind it. Nuuk's involvement caused consternation among other Inuit groups who resisted the concentration of power in the sovereign hands of the sub-Arctic capitals (Inuit Circumpolar Council 2009; Steinberg et al. 2014). Indigenous and postcolonial identity was thus a paradiplomatic resource for Greenland, which had implications for the high politics of the region, but it remained contested as other actors challenged Greenland's right to speak on behalf of the Inuit.

The Ilulissat Declaration illustrates how studying Greenland's relationship to the outside world improves our understanding of paradiplomacy in general, Arctic politics, and Greenlandic identity. First, Greenland illuminates hitherto unexamined aspects of modern paradiplomacy by highlighting how postcolonialism and high politics can affect a regional entity's paradiplomatic efforts. Instigated in the 1970s, the study of paradiplomacy has flourished since the 1990s, as European regions began to use diplomatic relations to the EU to gain leverage vis-à-vis their central governments, while Canada experienced a new wave of Québécois separatism (Keating 1999; Kuznetsov 2014). However, the paradiplomacy literature has mainly focused on regions within developed countries – American states, Canadian provinces (Quebec in particular), and European regions (especially the Basque region and Catalonia in Spain, and Flanders and Wallonia in Belgium) – whose main relations to foreign entities revolve around low politics issues, such as cultural cooperation and local economic development (Soldatos and Michelmann 1992; Aldecoa and Keating 1999; Lecours 2008 – exceptions include Sharafutdinova 2003; Bartmann 2006). Greenland differs from these cases in two crucial ways. First, Greenland is a geographically separate polity with a postcolonial relationship to the Danish state and an indigenous identity. As the abovementioned Ilulissat Declaration example shows, postcolonialism gives Greenland a strong paradiplomatic tool, but it also opens a new arena for paradiplomatic activity, as Greenland is forced to engage and compete with Inuit organizations for the right to represent the Inuit cause. Second, Greenland's paradiplomatic efforts also touch high politics issues that involve foreign great powers, such as the United States' geostrategic interests in the High North, regional governance structures of the Arctic, and Chinese strategic resource interests. By high politics, we mean policymaking that focuses on military, economic, and political interests that are essential for the long-term security of the states involved. As the Arctic opens up for new activity, more states will come to get crucial interests in the region and this will affect Greenland's diplomatic opportunities. High politics issues raise Greenland's profile on the agenda of larger states, but also pose specific challenges that differ from those that follow from low politics.

Second, the bulk of the literature on the international relations of the Arctic has largely focused on the importance of states and international institutions and has largely ignored the role that sub-state actors, including Inuit groups and entities, play in regional politics (e.g. Blunden 2009; Brosnan et al. 2011; Haftendorn 2011; Carlsson and Granholm 2013; Le Miere and Mazo 2013; Offerdal 2014 – exceptions include Joenniemi and Sergunin 2014; Powell and Dodds 2014; Shadian 2014). The literature on Danish Arctic policy, for instance, has largely refrained from examining Greenland as a foreign policy actor and has instead viewed it as an arena for Denmark's foreign policy, effectively silencing any indigenous foreign policy significance (Jørgensen and Rahbek-Clemmensen 2009; Petersen 2009; Degeorges 2012; Rahbek-Clemmensen et al. 2012; Wang and Degeorges 2014). Greenland is poised to play a more significant role in regional politics as it carves out an increasingly independent foreign policy and understanding the dynamics that drive Nuuk is therefore essential for Arctic analysts. Studying Greenland also opens the international relations literature to the importance of Inuit identity and the politics of sub-state actors. The present volume presents a myriad of processes and factors through which Nuuk affects the politics of the region and it points out that Greenland's position is informed and shaped by its particular Inuit identity. It functions a lens for understanding the effects that sub-state and Inuit groups have on regional politics and the volume thus illustrates how statehood is often fluid in the Arctic.

Finally, the Greenlandic politics literature lacks a comprehensive analysis of how Greenland, as an autonomous entity within the Danish realm, thinks about, conducts, and is simultaneously empowered and challenged by the new international relations of a changing Arctic. The main strands of the literature on Greenlandic politics and society has, often with an ethnographical micro-perspective, focused primarily on unpacking Greenlandic cultural identity, Danish colonial practices, or the independence process as such (e.g. Nuttall 1998; Christiansen 2000; Nuttall 2001; Thisted 2005; Graugaard 2009; Dahl 2010; Hastrup 2015; Sejersen 2015; Rud 2017). This important work has, however, come at the expense of a thorough engagement with the political significance of international relations. Accordingly, only a few scattered pieces have so far analysed how Greenland navigates the international realm and even fewer have examined how the island's relationship to external actors plays a role in shaping politics and identity (e.g. Petersen 2006; Gerhardt 2011; Gad 2014; Steinberg et al. 2014;Ackrén and Jakobsen 2015; Kleist 2016). As the present volume shows, ideas about independence and sovereignty play a crucial role for Greenlandic identity and acting like a state in an international realm of states is one key identity mechanism used by policymakers in Nuuk.

This volume examines how postcolonialism and high politics interests help carve room for paradiplomacy by exploring the intersection between Greenland and the international politics of the Arctic. Following Lecours and Moreno, it views paradiplomacy as a dialectical relationship, where a regional actor's paradiplomacy not only gives it new sources of power, but also help it to define its regional interests and to construct local national identities (Lecours and Moreno 2001). Regional identity and politics become intertwined with international politics, as paradiplomatic relations with foreign entities help shape local identity, while simultaneously playing a role in the international sphere. Existing studies of Greenland's paradiplomacy have focused on mapping specific, formal initiatives by the government of Greenland, but one has to go further than that to explore how the international realm plays a role in Greenlandic identity, how Greenland uses informal influence to shape foreign entities, and how foreign states try to navigate this space (Ackrén 2014a; 2014b). Consequently, the volume explores how international politics and external actors (including Denmark) affect Greenlandic politics and identity and, conversely, what role Greenland play in the foreign policies of the Arctic.

Greenland enters international politics

Greenland's current political relations and Greenlandic identity are both shaped by a history of colonialism and contested sovereignty. Greenland, the world's largest island and home to some of the world's most northern settlements, had been inhabited by people for millennia before the first Europeans, Icelandic and Norwegian Vikings, came to its shores in the late tenth century. Originally a Norwegian territory, Greenland came under Copenhagen when Denmark and Norway formed a personal union in fourteenth century and it stayed under the Danish crown when the dual monarchy was absolved with the Treaty of Kiel after the Napoleonic wars. The harsh climate and long distances ultimately isolated the first European settlers, who are thought to have succumbed to winter and competition from Inuit tribes during the Middle Ages, and the island was only brought permanently back into the European system of states, when Danish–Norwegian missionaries returned in 1721 (Gad 1967). Essentially a hub for missionaries and fishing and whaling for centuries, Danish explorers traversed the island and began to document the customs of the Inuit peoples in the late nineteenth and early twentieth centuries. Local district councils were established in the mid-nineteenth century, but real authority was kept in Copenhagen (Gad 1984, 211–74).[1]

Copenhagen struggled to get other states, most importantly the United States and the United Kingdom, to fully recognize its sovereignty over the island well into the twentieth century. During the First World War, American policymakers debated whether to buy Greenland only to decide against it. Instead, the American government bought the Danish West Indies and in a colonial *quid pro quo* issued a statement that "the government of the United States of America will not object to the Danish government extending their political and economic interests to the whole of Greenland" (Berry 2016, 108–109). Between the two world wars, following disputes between Danish and Norwegian hunters, Norway challenged Danish sovereignty over eastern Greenland, only to lose the dispute at the International Court in The Hague (Beukel et al. 2010, 19–20). During the Second World War, Danish diplomats allowed an American presence on the island in return for a vague recognition of Danish sovereignty and a Danish contribution to the Allied cause. After an unsuccessful American bid to buy Greenland in 1946, Washington and Copenhagen negotiated a defence agreement in 1951 that formalized a permanent American military presence (Government of Denmark and Government of the United States of America 1951; Danish Foreign Policy Institute 1997). The defence agreement *de facto* created a system of shared sovereignty over Greenland, where the United States military was allowed to operate whenever and wherever it needed, while staying out of Greenlandic affairs and allowing the Danes *de jure* sovereignty over the island (Lidegaard 1996, 574–86). Greenland came to play a central geostrategic role during the Cold

War and allowing the American presence boosted Denmark's otherwise limited contribution to NATO (Lidegaard 1996, 574–86; Villaume 1997; Ringsmose 2008, 211, 231). Denmark took the blame for several scandals that surfaced during or after the Cold War as a consequence of the American activities, including the presence of American nuclear weapons in Thule in the 1960s in contradiction with official Danish policy and the forced relocation of indigenous settlements from Thule in 1953 (to make room for the expanded American base).

As Danish sovereignty was secured and the threat of a foreign takeover dissipated, an internal source of uncertainty appeared. De-colonization and a rising national and political awareness among Greenlanders meant that independence became an issue and that the Danish government had to take Greenlandic interests into account. Greenland was de-colonized in 1953 and became a county of metropolitan Denmark, only to get home rule (semi-autonomy) in 1979 (Gad 1984; Beukel et al. 2010). Though this arrangement still located final authority over many issues, including the island's foreign and security policy, in Copenhagen, Nuuk could use its new-found status to carve out room for an independent foreign policy. Home rule thus allowed Greenland to leave the European Economic Community in 1985, to open its own representations in foreign capitals, and to get a seat at the table when the Thule base agreement was renegotiated in 2003 (Kristensen 2005). For a while, Greenlandic premiers even represented the Kingdom of Denmark in the Arctic Council (Arctic Council 2016).[2] Greenland loosened its ties to Copenhagen further in 2009 as a new Self-Government Act gave it a road-map to full independence and the possibility of acquiring areas of responsibility insofar as it could finance them (except foreign and security policy, which remained within Denmark's purview). The act recognized the Greenlanders as a people and it gave Greenland formal right to pursue an independent foreign policy in policy areas which fell under Greenland's purview, as long as it did not contradict the overall foreign policy of the kingdom. The Danish parliament continued to hold legislative power over issue areas that Greenland had not acquired, but was now required to hear the government of Greenland, but not the Greenlandic parliament (Government of Denmark 2009, Kleist 2016).

Greenland still depends on Denmark, as the annual block grant from the Danish state and transfers from the EU still make up 35 per cent of the Greenlandic GDP (Economic Council of Greenland 2014, 20). The opening of the Arctic also raises the possibility of new industrial opportunities in Greenland, including oil and gas exploration, mineral extraction, tourism, and hydroelectric energy projects, and thus also stokes the dream of a fully independent Greenland as well as increased foreign interest in the island. Several foreign companies are investing in Greenlandic resource extraction and many governments recognize the economic, political, and strategic potential in improved relations with Nuuk. With the dream of self-determination, however, comes tensions between Nuuk and Copenhagen, as the two governments disagree about who has authority over which policy area. Talk of Greenlandic secession and consequent tensions have dampened recently, as many projects have folded and economic prospects have soured. Discussions about authority are, however, always ongoing and it seems only a matter of time before heated debates about Greenlandic independence will resurface.

Structure of the volume

The volume is structured to highlight how paradiplomacy affects both local identities and international politics. Examining the complex intertwined nature of the domestic and the international entails exploring both identity and means/end strategic behaviour. The chapters therefore represent several disciplines and perspectives, and focus on different levels and areas. Though each contribution thus focuses on only a few aspects of Greenlandic foreign policymaking, they form an overall mosaic when read together that gives an overview of the dynamics that shapes Greenland's place in the world.

The volume begins by exploring how foreign relations and paradiplomacy play an important internal role by shaping local interests and identities by creating an Other against which Greenlandic identity can be defined. The postcolonial relationship to Denmark continues to affect Greenlandic identity, but as Greenland develops relations with other actors, they also come to play a role in the domestic setting. Marc Jacobsen and Ulrik Pram Gad thus open the volume in Chapter 1 by setting the scene of external actors that are brought into Greenlandic debates and showing how such Others affect Greenlandic self-conceptualization. Though Denmark have previously been the main foreign Other against which Greenlandic identity was built, in recent years other actors – including the US, China, and the EU, but also other Inuit and Arctic small states such as Iceland – and the international realm as such has come to play a more significant role. In Chapter 2, Jens Heinrich places this othering in a historical context by showing how local elites came to see international actors as crucial for Greenland and that the island's relation to the outside world came to shape the independence project from the beginning of the Second World War to Greenland left the European Community in 1985. These identity formations in turn affect Greenland's foreign relations. Kristian Søby Kristensen and Jon Rahbek-Clemmensen in Chapter 3 examine how the independence project currently plays out in concrete policy debates and how three separate visions – a political vision, an economic vision, and an environmental vision – shape how Greenland approaches the world. The enhanced importance of the Arctic relations that alters Greenland's position also affects how Denmark approaches the region. In Chapter 4, Jon Rahbek-Clemmensen shows that Danish thinking about Greenland has undergone an Arctic turn, whereby Danish policymakers may have directed their attention northward, but they have simultaneously become more interested in the politics of the Arctic region than the bilateral relationship to Greenland. Thus, the rising importance of the High North does not only strengthen Greenland's profile vis-à-vis external actors, it also reframes the region in Copenhagen in a way that may diminish Greenland's room for manoeuvre.

Turning to Greenland's actual external relations, it becomes clear that postcolonialism, the changing Arctic, and the complex constitutional set-up within the Kingdom of Denmark enable Nuuk to forge relations with outside actors, but that these factors simultaneously inhibit Greenland's room for manoeuvre. Mikkel Runge Olesen in Chapter 5 shows that reputation affects the triangular relationship between Greenland, Denmark, and the United States and that Copenhagen functions as a lightning rod that eases US–Greenlandic relations by shielding Washington from postcolonial blame. The paradiplomatic challenges involved in high politics become obvious, when one turns to Sino–Danish–Greenlandic relations. In Chapter 6, Camilla T. N. Sørensen argues that fears that China would take over Greenland are exaggerated, but that Danish and American concerns about China's future role in the region and the complex constitutional relationship within the Kingdom of Denmark have contributed to discouraging Beijing from enhancing its involvement in Greenland. Kevin Foley in Chapter 7 shows that Copenhagen's position has been driven by Danish domestic politics, as exaggerated concerns over national security and labour regulation played into the tactical manoeuvring of the major parties in parliament. This underlines that Greenland remains dependent on Danish domestic politics.

Sustaining a postcolonial position also entails maintaining an amenable relationship to other Inuit actors, who pursue other political objectives. Examining the relationship between Greenland and the Inuit Circumpolar Conference (ICC), Hannes Gerhardt demonstrates in Chapter 8 how each of these entities represents a different vision of Inuit politics. While the ICC problematizes a state-centric, Westphalian order in the Arctic, Greenland, the only Inuit community with a realistic path to statehood, pushes an Inuit counter-narrative that affirms the importance of states. In Chapter 9, Inuuteq Holm Olsen and Jessica M. Shadian use the same schema to examine Greenland's role in the Arctic Council. They show that the increasing importance of states in the Council marginalizes Nuuk from central decision-making and they thus highlight that the downsides of Greenland's acceptance of a state-centric High North order. Klaus Dodds and Mark Nuttall in Chapter 10 conduct a critical geopolitics analysis of Greenland's role in Arctic geopolitical discourses to show how our understanding of Greenland is shaped by geographical and spatial imaginaries about the surface and subsurface of the island and how these imaginations have changed since the Cold War. These geopolitical imaginations produce, in turn, the spatially defined conditions of possibility for independent Greenlandic politics and political contestation.

Notes

1   See also Jens Heinrich's contribution to the present volume, Chapter 2.
2   See also Jon Rahbek-Clemmensen's contribution to the present volume, Chapter 4.

References

Ackrén, M. (2014a). Greenlandic Paradiplomatic Relations. In: L. Heininen, ed., *Security and Sovereignty in the North Atlantic*. Basingstoke: Palgrave Macmillan, pp. 42–61.

Ackrén, M. (2014b). Paradiplomacy in Greenland. In: A. Grydehøj, L. Fabiani, J. S. I. Ferrando, L. Aristidi, M. Ackrén, eds, *Paradiplomacy*. Brussels: Centre Maurits Coppieters, pp. 64–78.

Ackrén, M. and Jakobsen, U. (2015). Greenland as a Self-Governing Sub-National Territory in International Relations: Past, Current and Future Perspectives. *Polar Record* 51(4), pp. 404–412.

Aldecoa, F. and Keating, M. (1999). *Paradiplomacy in Action: The Foreign Relations of Subnational Governments*. Abingdon: Routledge.

Arctic Council. (2016). All Arctic Council Declarations 1996-2015. Available from: https://oaarchive.arctic-council.org/bitstream/handle/11374/94/EDOCS-1200-v3-All_Arctic_Council_Declarations_1996-2015_Searchable.PDF?sequence=4&isAllowed=y. [Accessed 7 November 2016].

Bartmann, B. (2006). In or Out: Sub-National Island Jurisdictions and the Antechamber of Para-Diplomacy. *The Round Table* 95(386), pp. 541–559.

Berry, D. (2016). The Monroe Doctrine and the Governance of Greenland's Security. In: D. Berry, N. Bowles, H. Jones, eds, *Governing the North American Arctic: Sovereignty, Security, and Institutions*. Basingstoke: Palgrave Macmillan, pp. 103–121.

Beukel, E., Jensen, F., and Rytter, J. (2010). *Phasing Out the Colonial Status of Greenland, 1945–54: A Historical Study*. Copenhagen: Museum Tusculanum Press.

Blunden, M. (2009). The New Problem of Arctic Stability. *Survival* 51(5), pp. 121–142.

Borgerson, S. (2008). Arctic Meltdown. *Foreign Affairs* 87(2), pp. 63–77.

Brosnan, I., Leschine, T., and Miles, E. (2011). Cooperation or Conflict in a Changing Arctic? *Ocean Development & International Law* 42(1–2), pp. 173–210.

Carlsson, M. and Granholm, N. (2013). *Russia and the Arctic: Analysis and Discussion of Russian Strategies*. Stockholm: Swedish Defence Research Agency.

Chivers, C. (3 August 2007). Eyeing Future Wealth, Russians Plant the Flag on the Arctic Seabed, Below the Polar Cap. *The New York Times*. p. 8.

Christiansen, H. (2000). Arktisk symbolsk politik – eller hvorledes skal vi forstå den grønlandske idé om selvstændighed? *Politica* 32(1), pp. 61–70.

Dahl, J. (2010). Identity, Urbanization and Political Demography in Greenland. *Acta Borealia* 27(2), pp. 125–140.

Danish Foreign Policy Institute. (1997). *Grønland under den Kolde Krig, Dansk og Amerikansk Sikkerhedspolitik 1945–68*. Copenhagen: Danish Foreign Policy Institute.

Degeorges, D. (2012). *The Role of Greenland in the Arctic*. Paris: L'Institut de Recherche Stratégique de l'Ecole Militaire.

Economic Council of Greenland. (2014). *Grønlands Økonomi 2014*. Nuuk: Government of Greenland.

Gad, F. (1967). *Grønlands Historie*. Copenhagen: Nyt Nordisk Forlag.

Gad, F. (1984). *Grønland*. Copenhagen: Politikens Forlag.

Gad, U. (2014). Greenland: A Post-Danish Sovereign Nation State in the Making. *Cooperation and Conflict* 49(1), pp. 98–118.

Gerhardt, H. (2011). The Inuit and Sovereignty: The Case of the Inuit Circumpolar Conference and Greenland. *Tidsskriftet Politik* 14(1), pp. 6–14.

Government of Denmark. (2009). *Lov om Grønlands Selvstyre*. Lovtidende A, no 473, 13 June.

Government of Denmark and Government of the United States of America. (1951). *Forsvarsaftalen af 1951*. Available from: https://www.retsinformation.dk/Forms/R0710.aspx?id=70554 [Accessed 23 May 2017].

Graugaard, N. (2009). *National Identity in Greenland in the Age of Self-Government*. Working Paper. Peterborough: Trent University.

Haftendorn, H. (2011). NATO and the Arctic: Is the Atlantic Alliance a Cold War Relic in a Peaceful Region Now Faced with Non-military Challenges? *European Security* 20(3), pp. 337–61.

Hastrup, K. (2015). *Thule på Tidens Rand*. Copenhagen: Lindhardt og Ringhof.

Hertell, H. (2008). Arctic Melt: The Tipping Point for an Arctic Treaty. *Georgetown International Environmental Law Review* 21, pp. 565–591.

Inuit Circumpolar Council. (2009). A Circumpolar Inuit Declaration on Sovereignty in the Arctic. Available at: http://inuit.org/icc-greenland/icc-declarations/sovereignty-declaration-2009/. [Accessed 21 May 2017].

Joenniemi, P. and Sergunin, A. (2014). Paradiplomacy as a Capacity-Building Strategy. *Problems of Post-Communism* 61(6), pp. 18–33.

Jørgensen, H. and Rahbek-Clemmensen, J. (2009). *Keep It Cool! Four Scenarios for the Danish Armed Forces in Greenland in 2030.* Copenhagen: Danish Institute for Military Studies.

Keating, M. (1999). Regions and International Affairs: Motives, Opportunities and Strategies. *Regional & Federal Studies* 9(1), pp. 1–16.

Kleist, M. (2016). Greenland Self-Government and the Arctic. In: D. A. Berry, N. Bowles, H. Jones, eds, *Governing the North American Arctic: Sovereignty, Security, and Institutions.* Basingstoke: Palgrave Macmillan, pp. 247–252.

Koivurova, T. (2008). Alternatives for an Arctic Treaty – Evaluation and a New Proposal. *Review of European Community & International Environmental Law* 17(1), pp. 14–26.

Kristensen, K. (2005). Negotiating Base Rights for Missile Defence: The Case of Thule Air Base in Greenland. In: *Missile Defence: International, Regional and National Implications.* Abingdon: Routledge, pp. 183–207.

Kuznetsov, A. (2014). *Theory and Practice of Paradiplomacy: Subnational Governments in International Affairs.* Abingdon: Routledge.

Le Miere, C. and Mazo, J. (2013). *Arctic Opening: Insecurity and Opportunity.* London: International Institute for Strategic Studies.

Lecours, A. (2008). *Political Issues of Paradiplomacy: Lessons from the Developed World.* The Hague: Netherlands Institute of International Relations 'Clingendael'.

Lecours, A. and Moreno, L. (2001). *Paradiplomacy and Stateless Nations: A Reference to the Basque Country.* UPC Working Paper. Madrid: Instituto de Políticas y Bienes Públicos.

Lidegaard, B. (1996). *I Kongens Navn: Henrik Kauffmann i Dansk Diplomati 1919–1958.* Copenhagen: Samleren.

Nuttall, M. (1998). States and Categories: Indigenous Models of Personhood in Northwest Greenland. In: R. Jenkins, ed., *Questions of Competence. Culture, Classification and Intellectual Disability.* Cambridge: Cambridge University Press, pp. 176–93.

Nuttall, M. (2001). Locality, Identity and Memory in South Greenland. *Études/Inuit/Studies* 25(1/2), pp. 53–72.

Offerdal, K. (2014). Interstate Relations: The Complexities of Arctic Politics. In: R. Tamnes and K. Offerdal, eds., *Geopolitics and Security in the Arctic: Regional Dynamics in a Global World.* Abingdon: Routledge, pp. 73–96.

Petersen, H. (2006). *Grønland i Verdenssamfundet: Udvikling og Forandring af Normer og Praksis.* Nuuk: Forlaget Atuagkat.

Petersen, N. (2009). The Arctic as a New Arena for Danish Foreign Policy: The Ilulissat Initiative and its Implications. In: H. Mouritzen and N. Hvidt, eds, *Danish Foreign Policy Yearbook 2009.* Copenhagen: Danish Institute for International Studies, pp. 35–78.

Powell, R. and Dodds, K. (2014). *Polar Geopolitics?: Knowledges, Resources and Legal Regimes.* Cheltenham: Edward Elgar Publishing.

Rahbek-Clemmensen, J., Larsen, E., and Rasmussen, M. (2012). *Forsvaret i Arktis: Suverænitet, Samarbejde og Sikkerhed.* Copenhagen: Center for Military Studies.

Ringsmose, J. (2008). *Frihedens Assurancepræmie: Danmark, NATO og forsvarsbudgetterne.* Odense: Syddansk Universitetsforlag.

Rud, S. (2017). *Colonialism in Greenland: Tradition, Governance and Legacy.* Basingstoke: Palgrave Macmillan.

Sejersen, F. (2015). *Rethinking Greenland and the Arctic in the Era of Climate Change: New Northern Horizons*. London: Routledge.

Shadian, J. (2014). *The Politics of Arctic Sovereignty: Oil, Ice, and Inuit Governance*. London: Routledge.

Sharafutdinova, G. (2003). Paradiplomacy in the Russian Regions: Tatarstan's Search for Statehood. *Europe–Asia Studies* 55(4), pp. 613–629.

Soldatos, P. and Michelmann, H. (1992). Subnational Units' Paradiplomacy in the Context of European Integration. *Journal of European Integration* 15(2–3), pp. 129–134.

Steinberg, P., Tasch, J., and Gerhardt, H. (2014). *Contesting the Arctic: Rethinking Politics in the Circumpolar North*. London: I.B.Tauris.

The Ilulissat Declaration (2008). Available at http://www.oceanlaw.org/downloads/arctic/Ilulissat_Declaration.pdf. [Accessed 20 January 2015].

Thisted, K. (2005). Postkolonialisme i Nordisk Perspektiv. In: H. Bech and A. Sørensen, eds, *Kultur På Kryds Og Tværs*. Copenhagen: Klim, pp. 16–43.

Villaume, P. (1997). Henrik Kauffmann, den kolde krig og de falske toner. *Historisk Tidsskrift* 97(2), pp. 491–511.

Wang, N. and Degeorges, D. (2014). *Greenland and the New Arctic: Political and Security Implications of a State-Building Project*. Copenhagen: Royal Danish Defence College.

1 Setting the scene in Nuuk

Introducing the cast of characters in Greenlandic foreign policy narratives

Marc Jacobsen and Ulrik Pram Gad

Greenland's post-colonial foreign relations: diversification of the constitutive dependency

Greenland has for decades worked towards enhanced independent agency in international politics as a way to escape the unilateral dependency on Denmark. The renewed global interest in the Arctic has given new impetus to a strategy of diversifying its dependency relations as a way to post-coloniality. As the government of Greenland puts it in its foreign policy strategy; "It is important that the interest in the Arctic and Greenland is converted into concrete opportunities for the Greenlandic people and its development as a nation" (Government of Greenland 2011, 3). By referring to narratives of tradition and modernity, Greenland has used this increased interest in the Arctic to enhance relations with Inuit kinsmen, Nordic siblings, the UN, the USA and the EU while seeking to establish new bilateral relations with Asian powers. This chapter will put each of these relations in a historical perspective and investigate how Greenland's foreign policy is guided by the national self-image in combining symbolic elements of indigenous cultural traditions with envisioned future independence. The basic narratives of tradition and modernity, however, sometimes clash. Hence, the chapter introduces the core cast of characters in the most important narratives, which Greenland is telling about its place in the world through official documents, speeches and media statements, supplemented with secondary literature.[1] Empirically, the analysis, thus, gives priority to the collective narrative told by official Greenlandic representatives on the international level.

Theoretically, the analysis draws on the tradition[2] of analysing international politics and foreign policy as driven by narratively structured discourses constructing nation state identities in relation to different Others. If there was no difference, one could not meaningfully talk about identity. On the one hand, any identity needs a radical Other to exist (Derrida 1988, 52; Connolly 1991, 64f; Campbell 1998, ix–x). In Greenland, Denmark has for centuries been this central Other, at once constituting and threatening to eradicate Greenlandic identity. The emergence of collective Inuit identity seems to have been provoked by the encounter with *qallunaat* [white people], and in the case of Greenland, Danes were the *qallunaat* who stayed to make a lasting impression (Sørensen

1994, 109). This contrast is still defining for Greenlandic identity, as noted by Sørensen when concluding on his fieldwork in the housing projects of Nuuk: "[G]reenlandicness and Danishness are mutually experienced and applied as mutually negating each other in this ethno-political universe." (Sørensen 1991, 48) Denmark appears in Greenlandic identity discourse as those who first corrupted indigenous Greenlandic culture and identity (Gad 2005, 66ff; 2016, 46): To be authentic, Greenland needs a population which speaks Greenlandic, it needs hunters who provide for themselves by providing the Greenlanders with *kalaalimerngit* [Greenlandic food] and selling sealskin to *qallunaat*. However, this basic narrative of decline of traditional Inuit culture coexist unhappily with the inclusion of a series of distinctly modern elements in Greenlandic everyday life and identity discourse: No one imagines a Greenland which does not include 100hp outboard motors, the internet, Canadian Goose outdoor gear or democracy and "Scandinavian level" welfare services (Gad 2005). Hence, the narrative of decline of tradition has – each and every day – to be reconciled with a narrative of modernisation. In the combination of these two narratives, Denmark is cast as the one preventing the resurrection of Greenlandic identity in the form of an independent nation state (Gad 2005, 46f).

On the other hand, identity narratives seldom just relate the identity of the self to one other – most often a whole cast of characters is involved (Ricoeur 1988, 248; Hansen 2006, 40; Gad 2010, 38, 418). Over the decades since the instigation of home rule in 1979, the government of Greenland has increasingly engaged in foreign relations, identifying forums and relations which would allow Greenland to participate either separately or as part of a Danish delegation (Petersen 2006). At times, Greenland seemed to pursue a rather indiscriminate approach. Initially, the main objective was to gain entrance – and thereby recognition – rather than any particular substantial interest (Nielsen 2001, 15). Gradually, a more considered and prioritised approach developed.[3] In these efforts, "Greenland has always been hedging its bets in relation to the international society and world society by playing both the national horse and the affiliation and cooperation with the indigenous peoples of the world." (Petersen 2006, 17). Hence, Greenlandic identity narratives have involved an ever wider cast of characters; some primarily linked to tradition, others to modernisation; some casted in positive terms, some in negative terms.

ICC, UN and Nunavut: partners in tradition – symbolism and ambiguous practices

One relation taken up well before home rule is the wider pan-Inuit identity. This other – or perhaps rather; secondary self – is linked to one of the central discursive elements of Greenlandic identity, i.e. the notion that traditional culture is defining. In popular discourse, Inuit in Canada, Alaska and Chukotka are often mentioned as kinsmen, connoting not just linguistic and cultural ties but also blood ties (Sørensen 1994, 125; cf. Dorais 1996, 30; Kleivan 1999, 103). Inuit identity, hence, is related to an aboriginal Greenlandic identity (Sejersen 1999, 131), which may – at the extremes – be narrated as a past relict hindering modernisation (Gad 2005) or, conversely, as a golden past which Greenlandic nationalism aims to resurrect (Gad 2005). This schism was explicit between the succeeding versions of the *Greenland Home Rule policy in the area of Indigenous People* drafted by the Subcommittee on Foreign and Security Policy of the unilateral Greenlandic Commission whose work led to the revised 2009 self-government arrangement (Arbejdsgruppen 2000; 2001; 2002): Those who would like Greenlanders to identify with a past Inuit community use the categories "indigenous" and "Inuit" as positive references, while those who would like Greenlanders to see themselves as "regular, modern people" distance themselves from the "indigenous people" identity (cf. Christiansen 2000, 67f). Inuit and indigenous identification was important for some of the leading figures of the Siumut party (established in 1977) as they initiated Greenlandic home rule, and it was part and parcel of the youth rebellion which formed the radical Inuit Ataqatigiit party (established in 1978). The status of the pan-Inuit identity in Greenland today might best be paralleled to the status of Nordic identity in Scandinavia (Dorais 1996, 30f; cf. Hansen 2002, 55, 57): A secondary "we" that attracts – in some individuals – positive emotions as it symbolises a romantic dream of a pure version of all the positive aspects of our presence freed of the corrupting influence from outside, combined with a certain strategic use, as it grants access to specific rights at a collective level such as whaling (Jacobsen 2015, 7–9) and resources at an individual level such as jobs and paid travel (Søbye 2013).

Greenlanders play an active part in the global movement of indigenous peoples concentrating on the United Nations Permanent Forum on Indigenous Issues (UNPFII) and United Nations Expert Mechanism on the Rights of Indigenous People (EMRIP). These forums are important arenas for colonised peoples seeking to challenge both the individual states and the very system of sovereign states by appropriating and "stretching" the language and tools of the self-same states (Lindroth 2011). Specifically, the government of Greenland is represented as part of the Danish delegation in the UNPFII and EMRIP. Here, Greenland enjoys a special position well exemplified by former premier Kuupik Kleist's speech shortly after the introduction of self-government. He described Greenland's new status as a "de facto implementation of the declaration of indigenous peoples' rights" (Government of Greenland 2010, 22; cf. Jacobsen 2015, 6). Moreover, he envisioned that "the experiences of Greenland's process can serve as inspiration for others of the world's indigenous peoples in their struggle for greater autonomy and in their development as a people" (Government of Greenland 2010, 7).

At the same time, Greenlanders are also present at the UNPFII and EMRIP via the Inuit Circumpolar Council[4] (ICC), which is part of the caucus of indigenous NGOs. The ICC is a transnational organisation, spanning four states, involving people in Canada, Alaska, Chukotka, and Greenland who identify as Inuit. This central organisational vessel of pan-Inuit identity – while remaining intellectually in opposition to the *qallunaat* states who colonised Inuit – at occasions articulate its identity in a much more convoluted way, which resists any clear cut distinction between state and indigenous identity. Challenging the indivisible sovereignties over the Arctic claimed by the Arctic states, the ICC in 2009 adopted a "Circumpolar Inuit Declaration on Sovereignty in the Arctic" which insisted that "The inextricable linkages between issues of sovereignty and sovereign rights in the Arctic and Inuit self-determination and other rights require states to accept the presence and role of Inuit as partners in the conduct of international relations in the Arctic." The ICC based their challenge on the observation that "Sovereignty is a contested concept, however, and does not have a fixed meaning".[5]

When including in the analysis not just the text but also the murky practices behind the declaration, its claim appears, however, to be based as much in strategic thinking as in an essentially different life-world. Probably the most spectacular example of the involvement of the Danish state with the ICC was when Crown Prince Frederik of Denmark was named the official patron of its 2010 General Assembly taking place in Nuuk. Organisationally, the ICC consists of four "member parties", each representing Inuit in one of the four states, and each organised according to the laws of these states. In the case of Greenland, membership counts a range of civil society organisations. However, Inatsisartut, the Parliament of Greenland, has also acceded to the ICC charter. As home rule – and later self-government – is territorially rather than ethnically defined, a handful of the members of the Greenlandic delegation to international ICC meetings are appointed by parliament, including representatives of the moderate Atassut (established 1978) and Demokraatit (established 2002) parties, which have been reluctant to define their political projects in ethnic terms. Moreover, the annual budget of Inatsisartut pays a substantial part of the annual expenses of the Greenlandic body of ICC. Since 2014, the annual economic support to ICC via the Finance Act has, however, been gradually reduced from 4.39 million DKK in 2014 to 3.0 million DKK in 2016 (Government of Greenland 2016, 244), leaving the impression that the Siumut coalition in this instance downgrades the affiliation with ICC in favour of more modern and state-like activities (cf. Gerhardt 2011; Strandsbjerg 2014; Jacobsen 2015). This impression is supported by the very few times ICC is mentioned in the annual foreign policy reports since 2009, where it only appears in connection with the Arctic Council's Sustainable Development Working Group and the work at the UN; especially regarding the UN's World Conference on Indigenous Peoples in 2014.

Inuit identity is most often articulated in international forums such as the above mentioned, but it is also the foundation for bilateral relations both focusing on traditional minority rights and modern industrial development. Across the Davis Strait, Canada established Nunavut as a separate territory in 1999 following a Land Claims Agreement signed with Inuit representatives in 1993. Greenland's Home Rule Government soon after established good formal relations to its new neighbour by signing a Memorandum of Intent with the purpose of generating close cooperation on a list of shared key interests (Okalik and Motzfeldt 2000) without involvement from Ottawa and Copenhagen (Krarup 2000). The shared colonial past and common goal of increased autonomy have since then united Greenland and Nunavut resulting in joint statements and agreements regarding both culturally important preservation and development of new opportunities. In line with concerns for tradition, the two governments have concluded agreements regarding strengthening of their respective Inuit languages (Government of Greenland 2015c, 73–74) and joint management and research of the polar bear (Government of Greenland 2015c, 64). In continuation hereof, Greenland and Nunavut have on several occasions issued common statements emphasising dissatisfaction with the EU's ban on seal product import (Governments of Greenland and Nunavut 2014; Government of Greenland 2015b). In line with the modernisation narrative, attempts to establish a direct flight connection between Greenland and Nunavut have been made, although a permanent route has not been sustainable (Government of Greenland 2015c, 19) and mining experiences have been exchanged. Most recently, a delegation of Greenlandic politicians went to Nunavut, Ottawa and Saskatchewan to learn about their experiences with uranium mining (Government of Greenland 2015a). In climate politics, both perspectives on tradition and development combines in a common quest for indigenous peoples' right to development, as stated by the two governments and ICC at COP21 in Paris (Government of Greenland et al. 2015).[6] Altogether, Nunavut and Greenland are not only partners in tradition but also in transition, and as in the Indigenous world movement, Greenland enjoys presenting itself as the more developed version of indigeneity also in relation to Canadian kinsmen (cf. Søbye 2013).

Norden, USA and the EU: allies in modernity – more or less threatening

Like the Inuit, Nordic identity plays a positive role in Greenlandic identity narratives. Often it is rhetorically mobilised as a lesser, more positive other compared to either aggressive Danish modernity or a globalised, American capitalism (Lynge 1999:43). Hence, Norden's international image as an homogenous, peaceful, successful and benevolent alternative to standard Western capitalism (Katzenstein 1996; Lawler 1997; Archer 2000, 109; Campbell et al. 2006; Kuisma 2007) is reflected in Greenland. Norden allows modernity to appear in more embracing, less dominating guises than when brought by the Danish other alone. As the Inuit, Norden is often presented within a family metaphor. However, it is not always clear that the members of the family enjoy equal status. This is mirrored in, i.e., the fact that representatives in Nordic fora from Greenland, the Faroe Islands and the Sami are not offered interpreting services at official meetings, meaning that they have to use one of the so-called "state bearing languages": Danish, Finnish, Icelandic, Norwegian and Swedish (cf. Jacobsen 2015, 7). The discontent with this differential treatment has been the starting point for establishing an alliance between Greenland, Åland and the Faroe Islands in 2012 that gives the three autonomous areas the authority to speak on behalf of each other (Jacobsen 2015, 7). Government of Greenland's foreign policy report of 2013 described this alliance as "a pivotal development of Greenland's foreign relations" (Government of Greenland 2013, 12; cf. Jacobsen 2015, 7). Related, an identity as "West Nordic" – embodied by a series of smaller fora, containing only Greenland, the Faroe Islands and Iceland – is articulated once in a while. Most often, common geography, common economic structure (fisheries and sheep farming) forms the basis of the narratives. But the shared history as Danish colonies at times seems to be a significant driver behind both the positive identification between the three countries,[7] but also the basis for arguing for Greenlandic and Faroese equality with the sovereign states in the Nordic fora.[8]

In Greenland as elsewhere, Norden is often represented as a culturally and politically less brutal contrast to yet another Other: the USA (Lynge 1999, 43; Qvist 2016; cf. Hansen 2002, 57; Adler-Nissen and Gad 2013). However, there is no agreement on what the role of the USA is in relation to Greenland. Immediately after the 9/11 attacks in 2001, the parliamentary debates of Inatsisartut to a large extent circled around emergency planning for a possible war. In one of these debates, Atassut chair Daniel Skifte brought his general point home by alerting to the fact that when, during WWII, the supply ships stopped coming from Denmark, the Americans stepped in as both providers and protectors. However, competing representations of the USA – and the Thule Air Base (established in 1951) in particular – involve casting it as the neighbourhood bully, as a threat (qua target for bombing in the case of war), and an everyday nuisance (since the base limits the locals' movement) (Lynge 2002; Gad forthcoming). Finally, the very existence of the base – sanctioned by Danish authorities without consulting Greenlanders (Lynge 2002; Gad forthcoming) – has been a symbol for the lack of recognition of Greenland as an actor in international politics.

Jørgen Taagholt concluded an article detailing the evolution and function of the Thule Air Base by encouraging "[t]he Thule hunters [to] declare with pride that they have contributed to securing world peace [by being removed from the base area, as] the balance of terror worked, and the base was an important element in this. So Greenland and the Thule hunters have suffered privations, which have benefitted mankind. A Third World War was avoided." (Taagholt 2002, 101–2) Such a narrative has absolutely no resonance in Greenlandic identity politics which is based on the understanding that we, the Inuit, are peaceful; war and military affairs are not our affairs; at most it is a problem imposed upon us from outside. Notable in this regard has been the near-total absence of Russia in Greenlandic foreign policy narratives (also noted by Nielsen 2001, 21), spare the routine exchange of fishing quotas in the northeast Atlantic. When Greenlandic politicians make (rare) demands for military investments in Greenland, arguments mostly relate to services provided for civil purposes (fisheries control, search and rescue, oil spill response, etc.) (cf. Hammond 2016).

However, during the late 1990s in the context of a planned upgrade of the Thule radar, the Home Rule Government repeatedly demanded that they be allowed to approach Washington directly without the detour via Copenhagen: A common iteration claimed that "If only we could talk directly to the Americans, they would recognise our legitimate claims, but the Danes will not let us". Since 2002, Greenlandic prime ministers and ministers of foreign affairs have participated in a series of meetings with the USA, culminating in the trilateral 2004 Igaliko agreement and the subsequent instigation of a "Joint Committee" mandated to promote cooperation between the USA and Greenland.[9] So, the

USA has increasingly played its part as recognisers of independent Greenlandic subjectivity. However, not much practical cooperation has come out of the Joint Committee. Rather, the main gain for Greenland from the negotiations has come in relation to Denmark as a result of post-colonial politics of embarrassment: The series of Thule incidents – including, most spectacularly, the 1953 relocation of the Inughuit and the crash of an air plane armed with hydrogen bombs in 1968 – serve as a constant reminder of Denmark not always playing the benevolent, protective role it claims. The mere threat of public shaming for colonial wrongs serves as a bargaining chip for Greenland; but in relation to Denmark rather than to the USA (Kristensen 2004). So apart from the Igaliku agreement – involving a measure of formal recognition from the USA – the substantial results of this strategy takes from Denmark and gives to Greenland: First, a (rare) official apology was extracted from the government of Denmark for the forceful relocation of the Inughuit. Second, a separate airport was built at Qaanaaq in 2001, partly paid for by the Danish state as a remedy for the forced removal of the Inughuit from Thule. They are now spared the helicopter tour to the USA base when going to other parts of Greenland or Denmark – provided that they can pay for a ticket in the first place.[10]

Finally, the consecution of Thule affairs also lay behind, the Danish state formally delegating the "full powers to the Government of Greenland to negotiate and conclude agreements under international law on behalf of the Kingdom of Denmark where such agreements relate solely to matters for which internal powers have been transferred to the Greenland Authorities" (Ministry of Foreign Affairs 2005).[11]

Alongside the US in the line-up of "other Others", the European Union has long taken up a special role in Greenlandic identity discourse. First, the casting of the EU, originally: the EEC, has, like the USA, been a point of contention in Greenlandic politics. Moreover, like the USA's entry on the scene, Danish entry into the EEC in 1973 has arguably been decisive for bringing Greenlandic aspirations a step towards self-government at a certain stage: contrary to the Faroes, who enjoyed home rule since 1948 and, hence, had a separate referendum, Greenland participated in the Danish referendum as any other provincial county. Under that constitutional status, the Greenlandic majority against membership did not matter. The experience of being included in the EEC against its will provided the impetus for finally demanding home rule (Gad 2016; Skydsbjerg 1999). This new status, modelled on the Faroese precedence, made retraction possible – and Greenland has since 1985 been able to praise itself with being the only "country" to ever leave the EU.

Moreover, the legal arrangement vis-à-vis the EU arrived upon quite early after the introduction of home rule established an independent subjectivity for Greenland. Contrary to the USA, in relation to whom Greenland only gained a direct voice after 2002, the EU has been conducive to Greenlandic agency for a longer period. Greenland utilises this platform for agency established long ago and has gradually expanded to further its position in the new opportunities by improved prospects for being able to harvest resources in the Arctic (cf. Gad et al. 2011; Gad 2016). The Association of the Overseas Countries and Territories of the European Union (OCTA)[12] is central to the positioning of Greenland as an individual international actor, taking up a leading role among the OCTA's members. Through this association, a number of potential allies are constructed, while both young Greenlandic diplomats and ministers use the forum as a place to learn the trade of international relations in practice with no superiors posing as superiors. Consequently, OCTA is revered as a nice place for Greenland to practice for sovereign equality (Gad 2016).

New Others from the East

During the first decades of the new millennium, new Others from the East have been appearing on the radar. In the beginning, the interest focused on China. In a parallel oscillation to the image of the USA, Greenlandic foreign policy narratives featured this new Other in diametrically opposing roles. In the late 1990s, the Chinese were cast as the saviours of the Greenlandic national trade, sealing, as a business development project (PUISI A/S), sponsored by the Home Rule Government, promised to turn surplus seal meat into cash by selling it as sausages in China. A few years later, Greenland in 2001 played a football match against Tibet as a way of showing sympathy and identification with another colonised people denied access to official FIFA tournaments. This narrative positioned China as an evil oppressor and the match triggered Chinese threats to the Greenlandic shrimp exports and cautious Home Rule Government efforts to defuse the problem (Mortensen 2007; Nybrandt and Mikkelsen 2016). However, after the introduction of self-government in 2009, the negative characterisation of the Chinese has faded and instead the role as economic saviours resurfaced – this time qua potential investments in minerals extraction in Greenland. But also officials from Japan and South Korea have become frequent guests in Greenland, enhancing the perception that the connections to the Far East are developing as still more viable trading alternatives to Europe, North America and the Nordic Countries. Particularly, hopes were high that Asian investments in mining would make full Greenlandic independence possible.

At the introduction of self-government on 21 June 2009, representatives from China, Japan and South Korea stood out as some of the more unusual official guests in Nuuk (Government of Greenland 2010, 33). Their presence was a sign of the renewed global interest in the Arctic, which, sparked by climate changes and their consequences, had become "a magnet for different countries' spheres of interest" (Government of Greenland 2009, viii). In the subsequent years, Greenland's rare earth elements became the centre for attention from China, the EU and later South Korea, which for some time competed to win Greenland's favour. It all accelerated in 2011, when then Minister for Industry and Labour, Ove Karl Berthelsen, paid an official visit to China where he met with number two in the hierarchy of China's government, Li Keqiang, who – as explained by Government of Greenland at home – showed a significant interest in Greenland's mineral potential (Government of Greenland 2012, 51). China's Minister for Land

and Resources, Xu Shaoshi, reciprocated the visit in April 2012 (Government of Greenland 2012, 53) and in response to these events, then EU Commissioner for Industry, Antonio Tajani, entered the fray and signed a Memorandum of Understanding (MoU) on future mining with Greenland (Government of Greenland 2013, 15). The simultaneous interest from two of the world's greatest powers only enhanced Greenland's bargaining position (cf. Kleist in Hansen 2012a) and when South Korea's president Lee Myung-Bak visited Greenland shortly after and signed a MoU regarding future mining cooperation (Government of Greenland 2014, 50), it was, as Premier Kleist underlined, "yet another proof of Greenland's enhanced foreign policy profile" (Hansen 2012b).

Since the football match in 2001, relations to the three big Asian powers had only been described in terms of economic importance. And when EU's interest in Greenland's mining potential did not materialise due to lack of finance, cultural understanding and ethnicity became part of the explanation, posing Asians as closer to Greenlanders. In an interview with the Danish newspaper *Weekendavisen*, Premier Kleist described Westerners' attitude as a master mentality with lack of respect for Greenlandic culture: In contrast, "East Asian businessmen apparently do make themselves more acquainted with the state of affairs than Western investors do, who apparently do not care where they are" (quoted in Andersen 2013). He also pointed out another, for him, natural community with the Asian partners: "Genetically, we Greenlanders are also, after all, more like family with people from the East" (quoted in Andersen 2013). The idea that Inuit's ancestors' historic migration patterns from Asia to the Arctic remains relevant, is not unique: Princess Alexandra, the first wife of Prince Joachim (Queen Margrethe of Denmark's younger son), has a Chinese father. When she donned a traditional Greenlandic women's costume during an official visit, the reaction on the streets was enthusiastic; "she looks just like us!". Nevertheless, the premier's weaving of genetics into the main storyline of Greenland's general foreign policy communication about the new Others from the East was new. Indeed, relations with Asia is almost exclusively described in economic terms, hence indirectly referring to the prospective discursive repertoire focusing on modernisation and enhanced self-government.

However, what in the beginning looked like a race between two of the world's most powerful regions, quickly became an exclusively Asian affair: the EU keeps buying fishing quotas and contributes to the "sustainable development of the education sector" via a partnership agreement – but European investment in mining never materialised. Relations in the Arctic Council evolved in a parallel way: China, Japan and South Korea were granted permanent observer status in 2013, while the EU's application was turned down by the Canadians explicitly as a way to protest against the EU's ban on import of seal products (Government of Greenland 2013, 1). Though the ban included a so-called "Inuit exception" (EU 2009), it *de facto* killed Greenland's seal product exports (Sommer 2012). In response to this development, Greenland has again looked towards the Far East, which is the "strongest fur market in the world" (Government of Greenland 2014, 26). While the interest in Greenland's mining potential has cooled down lately due to, i.a., the lower global market prices, more efforts have instead been invested in promoting Greenland's seal fur, seafood and tourist destinations to the three strong Asian economies. Most recently as part of an official visit by premier Kim Kielsen to Japan, which is currently Greenland's most important seafood export market outside the EU (Government of Greenland 2015c, 30-31). In this perspective – and with the need for investments in anticipated future mining projects in mind – Greenlandic politicians increasingly present relations with the Far East as crucial to Greenland's economic development and as a central way of diversifying dependency to the outside world.

Thus, for the past decade, the retrospective identification with the Tibetans as a colonised people has given way to economic considerations. Nevertheless, every now and then, single-minded focus on economic visions for the future is challenged. Sara Olsvig, then opposition leader, suggested that Greenland should send an official invitation to Dalai Lama to make clear that Greenland is in favour of human rights (EM2015/14 05:45:12-05:45:29). Vittus Qujaukitsoq, Minister of Foreign Affairs, replied that "It would be interesting if Dalai Lama from Tibet was invited to Greenland. I think that you in Inatsisartut must assess what is most important: trade, climate or human rights. What do you find most important?" (EM2015/14., 05:47:57-05:48:22). Qujaukitsoq's smile at that moment and the fact that Dalai Lama has still not been invited indicate that Greenland's international relations with the three Asian powers – China in particular – will continue focusing on business (as usual) in the years to come. Such prediction finds support in the fact that Qujaukitsoq cancelled a planned trip to Taiwan at the request of China (Karner and Damkjær 2017).

Conclusion: new opportunities brought by the "Arctic bonanza" discourse

Greenlandic identity narratives have involved an ever-wider cast of characters: some primarily linked to tradition, others to modernisation; some cast in positive terms, some in negative terms.[13] The very insistence on diversifying dependency relations from one relation (Copenhagen) to a variety of relations may be counted as one way of moving Greenlandic identity into a post-colonial mode, even if this version of post-colonialism does not (yet) involve full, formal sovereignty. One permanent is the often articulated wish for more independent control with external relations. Considering Asia, the government of Greenland claimed that Greenland getting in the driver's seat will "reduce any possible signal confusion considerably" (Government of Greenland 2014, 26). This endeavour has a clear precursor in the way in which the weight slowly being shifted from Copenhagen to Nuuk in the triangular relation between Greenland, Denmark and the USA: In 2013, Greenland got its own representation at the Danish embassy in Washington D.C. – just like the one established in Brussels in 1992. Following the same script, the establishment of a permanent representation in Beijing is often mentioned as the obvious next step (Government of Greenland 2014, 26). Like the USA and the EU, Asian countries are cast to play central roles as recognisers of independent

Greenlandic subjectivity, crucial for the process towards a more autonomous Greenland. In this regard, it might be counted as significant, that the Greenlandic representation in Ottawa established in 1998 was closed down already in 2002. In practice, Greenlandic politicians value global investments over cultural connections (cf. Strandsbjerg 2014).

However, Greenlandic foreign policy narratives exhibit a distinct affection when it comes to multilateral fora in which their representatives may pose in virtual sovereign equality: as mentioned above, Greenland takes pride of being first among equals in the Association of Overseas Countries and Territories (OCTA) in Brussels. In the West Nordic Council, Greenland gets to sit at the same table as sovereign Iceland – without Denmark. In the Arctic Council to the contrary, the government of Greenland is represented via the Kingdom of Denmark's delegation – a controversial arrangement in Greenland. Similar Greenlandic dissatisfaction has been articulated in relation to the way Denmark represents distinct Greenlandic interests in the World Trade Organisation and in the International Whaling Commission (Jacobsen 2015). However, the growing global attention and the fact that Denmark's presence in the Arctic is *only* legitimised by Greenland being a part of the Danish realm do make the Arctic Council a strategically well-chosen forum for articulating the wish for a more autonomous foreign policy.

In the Arctic Council, Greenlanders may be part of the ICC's delegation which enjoys the unique status of permanent participant qua indigenous people. Moreover, then premier Lars Emil Johansen did indeed sign the Ottawa declaration forming the Arctic Council in 1996 – but he did so representing Denmark by delegation. Greenland cannot speak in its own right but only as a representative for Denmark, and only when granted the right to do so by Denmark. In May 2013, Greenland's lower hierarchical status within this constellation was the centre for then premier Aleqa Hammond's boycott of the ministerial meeting in Kiruna, Sweden: if she could not have a Greenlandic flag at her own table in the first row, she preferred not to be part of the meeting (Josefsen 2015). The Greenlandic opposition met Hammond's tactics with harsh criticism, but the opposition, however, also shared the opinion that both self-governing territories and indigenous peoples should have a greater independent voice in the Arctic Council (Sørensen 2016). So in relation to the Arctic Council, Greenland seems to be hedging its bets even further by playing not just the national and the indigenous horse (cf. Petersen 2006, 17) – participating both through the Danish delegation and the ICC – but also fielding territorial and sub-regional horses: Greenland argues for including autonomous territories (like Greenland, Nunavut and Alaska) formally in the council's work. And to complicate matters further, the West Nordic Council – combining sovereign Iceland with autonomous Greenland and Faroes – is reportedly preparing an application for observer status in the Arctic Council (Veirum 2016). In this way, Greenland's representatives oscillate between emphasising either tradition or modernity, depending on whether it is a claim to Inuit identity or is state-centred geopolitics, which may open a door or improve Greenland's room for manoeuvre (cf. Jacobsen 2015).

Greenland has utilised its unique position as a very small population in command of a very large island on its way to break free from colonial subjugation, to gradually be evermore at the helm of its own foreign affairs on the verge of an Arctic bonanza. The initial tactics pursuing indiscriminate access and visibility has gradually been surpassed by a new tactics prioritizing fora and relations which specifically serve the purpose of diversifying dependence upon several sources of external resources. The different narratives used in this regard have both made the acquisition of full formal Greenlandic sovereignty more plausible and meanwhile it has widened the room to manoeuvre within imperial Danish sovereignty. Notably, Greenland has made this achievement by refraining from a general confrontation (rhetorical or otherwise) with its imperial metropole, Denmark. Rather, confrontations with the colonial overlords have been carefully calculated and occasionally staged to achieve maximum concessions from other Others like the USA and the EU. These calculations and stagings have, of course, relied on a particular constellation of past and future: the undeniable history of colonial subjugation combined with the enticing projection of an Arctic bonanza ahead. Greenland, hence, may credibly present itself as both a victim of past Danish colonisation and an important, more individual, player in an anticipated prosperous future Arctic in ways, which are not open to other micro-polities. However, as a general strategy for foreign relations even this new, more considered and prioritized tactics is so far an investment that has yet to produce returns beyond brand recognition.

Notably, even if squarely placed in the middle of the Arctic, Greenland's foreign policy narratives and initiatives hardly sum up to a fully-fledged "Arctic policy". However, elements in the initial visibility tactics and the new tactics of diversifying dependence may come to form the core of a third generation of narratives and initiatives aiming at strategically shaping the Arctic region as a preferred stage to act on for Greenland on its way to independence: First, since the establishment of the Arctic Council, a priority for Greenland has been to expand focus from environmental protection to allow for people actually living in the Arctic. In the AC, this ambition has condensed under the headline "sustainable development", allowing both care for indigenous culture and – particularly significant in Greenlandic politics – the exploitation of non-renewable resources (Gad et al. 2016). Second, the Greenlandic pragmatism in relation to "what horse to ride" – the indigenous, the autonomous, the sovereign – could coalesce into a principled promotion of the Arctic as a unique laboratory for creative governance structures: The real test for this ideal would, of course, be whether Greenland will still advocate the participation of indigenous peoples and autonomous territories when it flies its own flag in the front of the UN building in New York as a sovereign state.

Notes

1 The analysis presented here is updated and developed on the basis of thoughts presented in Gad (2005; 2016) and Jacobsen (2015). Most of the quotes referred to in

this chapter are originally in Danish or Greenlandic. The authors are responsible for the translation from Danish to English.

2 For more detailed presentations of (varieties of) the analytical strategy employed and its theoretical bases, cf. Jacobsen 2015; Gad 2005; 2010; 2016.

3 That is, via the introduction of cost/benefit analyses in the annual Foreign Policy Reports presented by the government of Greenland to the parliament. The Inuit Ataqatigiit (IA) led government took yet another step in this direction when it presented a Foreign Policy *Strategy* (Government of Greenland 2011).

4 Originally Inuit Circumpolar Conference, but re-named to signal its more permanent structure.

5 Available at http://inuit.org/fileadmin/user_upload/File/declarations/ICC_Sovereignty_ Declaration_2009_poster.pdf (accessed 13 March 2014); cf. Shadian (2010).

6 For more on the paradoxes of Greenlandic self-representation in relation to climate change, cf. Bjørst 2012.

7 The west-Nordic identity narratives (Thorleifsen, 2003) and organizations (Fjeldsbø, 2015) sometimes morphs eastwards to include all of coastal Norway (which enjoys a comparable economical relation to the north Atlantic fisheries and endures comparable geographical challenges) or just Finnmarken (which has a comparable history of colonial monopoly trade).

8 Cf. the press release from West Nordic Council following a joint meeting of the presidencies of West Nordic Council and Nordic Council; "Vestnordiske lande, ikke områder", posted 27 October 2009, available at http://www.norden.org/sv/aktuellt/ nyheter/vestnordiske-lande-ikke-omraader (accessed 13 March 2014).

9 Cf. the presentation on the website of the embassy of the USA to Denmark; http:// denmark.usembassy.gov/gl/jc.html (accessed 13 March 2014).

10 Seen from the local perspective of the inhabitants in Qaanaaq, the Home Rule/Danish sponsored disengagement from the USAF base is a much more ambiguous affair (cf. Gad forthcoming).

11 Except "Agreements affecting defence and security matters"; "Agreements which shall also apply to Denmark"; and "Agreements to be negotiated within an international organisation of which the Kingdom of Denmark is a member."

12 For a full list of OCTA members see: http://www.octassociation.org/octa-presentation

13 A recent opinion poll (Ugebrevet A4, 2013) reported 31 per cent of respondents in Greenland to prefer Canada (probably imagined primarily as the home of fellow Inuit) as the "closest *future* ally" (italics inserted) of Greenland, hence bypassing Denmark (22 per cent) and all other listed alternatives (the US, Norway, China and "Others" each preferred by less than 10 per cent); 25 per cent did not know how to answer. However, a different question in the same poll revealed that 84 per cent approved of "Greenland's participation in the Community of the Realm with Denmark and the Faroes" (for the time being, one supposes). These numbers are compatible on the background of earlier polls consistently indicating that 80–90 per cent favours independence – however, with 80–90 per cent adding that independence should not reduce the level of welfare (Skydsbjerg 2002).

References

Adler-Nissen, R. and Gad, U. (2013). Conclusion: When European and postcolonial studies meet. In: R. Adler-Nissen and U. Gad, eds., *European Integration and Postcolonial Sovereignty Games: The EU Overseas Countries and Territories*. London: Routledge, pp.235–245.

Andersen, M. (2013). Er kinesere værre end andre kapitalister?. *Weekendavisen*.

Arbejdsgruppen vedrørende Udenrigs- og Sikkerhedspolitik. (2000). *Grønlands Hjemmestyres politik på området Verdens Oprindelige Folk*. Arbejdspapir nr.2 til Selvstyrekommissionen.

Arbejdsgruppen vedrørende Udenrigs- og Sikkerhedspolitik. (2001). *Grønlands Hjemmestyres politik på området Verdens Oprindelige Folk.* Arbejdspapir nr.2 til Selvstyrekommissionen.

Arbejdsgruppen vedrørende Udenrigs- og Sikkerhedspolitik og Arbejdsgruppen vedrørende forfatningsretlige og folkeretlige spørgsmål. (2002). *Grønlands Hjemmestyres politik på området Verdens Oprindelige Folk.* Arbejdspapir nr.2 til Selvstyrekommissionen, rev.2, februar .

Archer, C. (2000). Euroscepticism in the Nordic region. *Journal of European Integration* 22(1), pp. 87–114.

Bjørst, L. (2012). Politiske positioner og skift i den grønlandske klimadebat fra 2001 til 2011. *Grønland* 60(1), pp. 2–19.

Campbell, D. (1998). *National Deconstruction: Violence, Identity, and Justice in Bosnia.* Minneapolis, MN: University of Minnesota Press.

Campbell, J., Hall, J. and Pedersen, O., eds. (2006). *National Identity and the Varieties of Capitalism: The Danish Experience.* Copenhagen: DJØF Publishing.

Christiansen, H. (2000). Arktisk symbolsk politik – eller hvorledes skal vi forstå den grønlandske idé om selvstændighed? *Politica* 32(1), pp. 61–70.

Connolly, W. (1991). *Identity\Difference: Democratic Negotiations of Political Paradox.* Ithaca, NY: Cornell University Press.

Danish Ministry of Foreign Affairs (2005). Circular note. File No. 8.U.107. Available at: http://www.stm.dk/multimedia/CirkularNote_GR.pdf. [Accessed 12 March 2015].

Derrida, J. (1988[1977]). Limited Inc a b c …In: J. Derrida, ed., *Limited Inc.* Evanston, IL: Northwestern University Press, pp 29–110.

Dorais, L.-J. (1996). Inuugatta inuulerpugut: Kalaallit and Canadian Inuit Identities. In: B. Jacobsen, ed., *Cultural and Social Research in Greenland 95/96: Essays in Honour of Robert Petersen.* Nuuk: Ilisimatusarfik/Atuakkiorfik, pp. 28–33.

EM2015/14 (2015). *Udenrigspolitisk redegørelse 2015.* Naalakkersuisoq for Erhverv, Arbejdsmarked, Handel og Udenrigsanliggender. Streaming available via: http://inatsisartut.gl/samlingerhome/oversigt-over-samlinger/samling/punktliste.aspx [Accessed 25 February 2016].

EU. (2009). *Europa-Parlamentets og Rådets Forordning (EF) Nr. 1007/2009 af 16. september 2009 om handel med sælprodukter (EØS-relevant tekst).* Den Europæiske Unions Tidende. L286/36.

Fjeldsbø, T. (2016). Forestillinger og Motforestillinger om Nord Atlanteren. En studie av NORA-regionen. Unpublished internship essay. Copenhagen: University of Copenhagen.

Gad, U. (2005). Dansksprogede grønlænderes plads i et Grønland under grønlandisering og modernisering. Copenhagen: Eskimologis Skrifter nr. 19.

Gad, U. (2010). (How) can They become like Us? Danish identity politics and the conflicts of 'Muslim relations'. PhD dissertation. Copenhagen: University of Copenhagen.

Gad, U. (2013). Greenland projecting sovereignty: Denmark protecting sovereignty away. In: R. Adler-Nissen and U. Gad, eds., *European Integration and Postcolonial Sovereignty Games: The EU Overseas Countries and Territories.* Abingdon: Routledge, pp. 217–34.

Gad, U. (2016). *National Identity Politics and Postcolonial Sovereignty Games: Greenland, Denmark, and the European Union.* Copenhagen: Museum Tusculanum Publishers.

Gad, U. (forthcoming). Pituffik in practice. Sex, lies, and airstrips, draft manuscript.

Gad, U., Hannibal, I., Holst, K. and Adler-Nissen, R. (2011). EUs oversøiske lande og territorier: postkoloniale suverænitetsspil og Grønlands arktiske muligheder. *Politik* 14(1), pp. 15–24.

Gad, U., Jacobsen, M., Graugaard, N., and Strandsbjerg, J. (2016). Politics of Sustainability in the Arctic: Postcoloniality, Nature, and Development. Paper presented at a workshop on 'Politics of Postcoloniality and Sustainability in the Arctic', Snekkersten, Denmark 27–28 May.

Gerhardt, H. (2011). The Inuit and sovereignty: The case of the Inuit Circumpolar Conference and Greenland. *Politik* 14(1), pp. 6–14.

Government of Greenland (2010). Udenrigspolitisk Redegørelse 2010. Available at: http://naalakkersuisut.gl/~/media/Nanoq/Files/Attached%20Files/Udenrigsdirektoratet/DK/Udenrigspolitiske%20redegorelser/Udenrigspolitiske%20redeg%C3%B8relse%202010.pdf [Accessed 15 October, 2015].

Government of Greenland (2011). Udenrigspolitisk Strategi 2011. Available at: http://naalakkersuisut.gl/~/media/Nanoq/Files/Attached%20Files/Udenrigsdirektoratet/DK/Udenrigspolitiske%20redegorelser/Udenrigspolitiske%20redeg%C3%B8relse%202011.pdf [Accessed 15 October 2015].

Government of Greenland (2012). Udenrigspolitisk Redegørelse 2012. Available at: http://naalakkersuisut.gl/~/media/Nanoq/Files/Attached%20Files/Udenrigsdirektoratet/DK/Udenrigspolitiske%20redegorelser/Udenrigspolitiske%20redeg%C3%B8relse%202012.pdf [Accessed 15 October 2015].

Government of Greenland (2013). Udenrigspolitisk Redegørelse 2013. Available at: http://www.ina.gl/media/1123614/pkt14_em2012_uspr_rg_dk.pdf. [Accessed 15 October 2015].

Government of Greenland (2014). Udenrigspolitisk Redegørelse 2014. Available at: http://www.landstinget.gl/dvd/EM2014/pdf/media/1901101/pkt14_em2014_udenrigspolitisk_redeg_relse_2014_dk.pdf. [Accessed 19 October 2015].

Government of Greenland (2015a). Nyttig uranviden indsamlet på oplysningstur til Canada. Available at: http://naalakkersuisut.gl/da/Naalakkersuisut/Nyheder/2015/06/urantur_canada [Accessed 29 February 2016).

Government of Greenland (2015b). Regeringerne i Nunavut og Grønland afgiver fælles erklæring. Available at: http://naalakkersuisut.gl/da/Naalakkersuisut/Nyheder/2015/04/faelleserklaering [Accessed 29 February 2016).

Government of Greenland (2015c). Udenrigspolitisk Redegørelse 2015. Available at: http://inatsisartut.gl/dvd/EM2015/pdf/media/2639392/pkt14_em2015_udenrigspolitiske_redeg_relse_dk.pdf. [Accessed 29 February 2016].

Government of Greenland (2016). *Finanslov for 2016.* Available at: http://naalakkersuisut.gl/~/media/Nanoq/Files/Attached%20Files/Finans/DK/Finanslov/2016/Finanslov%202016%20DK.pdf [Accessed 27 July 2016].

Governments of Greenland and Nunavut (2014). *Joint Statement of the Governments of Greenland and Nunavut regarding the Revision of the EU Seal Ban Regulation.* Available at: http://naalakkersuisut.gl/~/media/Nanoq/Files/Attached%20Files/Udenrigsdirektoratet/Joint%20statement%202015/240415_Joint%20Statement%20ENG.pdf. [Accessed 29 February 2016].

Government of Greenland, Government of Nunavut and Inuit Circumpolar Council (2015). *Governments of Nunavut and Greenland, and Inuit Circumpolar Council issue joint statement on climate change.* Available at: http://naalakkersuisut.gl/~/media/Nanoq/Files/Attached%20Files/Engelske-tekster/Aftaler/FINAL%20EN%20Joint%20statement%20on%20climate%20change.pdf [Accessed 29 February 2016].

Hammond, A. (2016). Danmark, giv Grønland en ny ordning. *Politiken.* Section 2, p. 7.

Hansen, L. (2002). Sustaining sovereignty: the Danish approach to Europe. In: L. Hansen and O. Wæver, eds., *European Integration and National Identity: The Challenge of the Nordic States,* Abingdon: Routledge, pp. 50–87.

Hansen, L. (2006): *Security as Practice. Discourse Analysis and the Bosnian War*. Abingdon: Routledge.

Hansen, M. (2012a). EU har giftet sig med Grønland. *Kalaallit Nunaata Radioa*. Available at: http://knr.gl/da/nyheder/eu-har-giftet-sig-med-gr%C3%B8nland. [Accessed 1 November 2015].

Hansen, M. (2012b). Underskrev aftale med Sydkorea. *Kalaallit Nunaata Radioa*. Available at: http://knr.gl/da/nyheder/underskrev-aftale-med-sydkorea. [Accessed 1 November 2015].

Jacobsen, M. (2015). The power of collective identity narration: Greenland's way to a more autonomous foreign policy. In L. Heininen, H. Exner-Pirot and J. Plouffe, eds., *Arctic Yearbook 2015: Arctic Governance and Governing*. Akureyri: Northern Research Forum, pp. 102–118.

Josefsen, L. (2013). Vi må gøre noget drastisk. *Sermitsiaq*. Available at: http://sermitsiaq.ag/node/154195 [Accessed 1 March 2016].

Karner, L. and Damkjær, O. (2017). Kina fik grønlandsk minister til at droppe Taiwan-besøg. *Berlingske*. Available at: http://www.b.dk/globalt/kina-fik-groenlandsk-minister-til-at-droppe-taiwan-besoeg [Accessed 26 January 2017].

Katzenstein, P.J. (1996). Regionalism in comparative perspective, *Cooperation & Conflict* 31(2), pp. 123–159.

Kleivan, I. (1999). Sprogdebatten. In: J. Lorentzen, E. Jensen and H. Gulløv, eds., *Inuit, kultur og samfund – en grundbog i eskimologi*. Aarhus: Systime.

Krarup, P. (2000). Udenrigspolitik på grønlandsk. *Kristeligt Dagblad*. Available at: http://www.kristeligt-dagblad.dk/kirke-tro/udenrigspolitik-p%C3%A5-gr%C3%B8nlandsk [Accessed 29 February 2016].

Kristensen, K. (2004). *Greenland, Denmark and the Debate on Missile Defense: A Window of Opportunity for Increased Autonomy*. Working Paper. Copenhagen: Danish Institute for International Studies.

Kuisma, M. (2007). Social democratic internationalism and the welfare state after the "Golden Age". *Cooperation and Conflict* 42(1), pp. 9–26.

Lawler, P. (1997). Scandinavian exceptionalism and European Union. *Journal of Common Market Studies* 35(4), pp. 565–94.

Lindroth, M. (2011). Paradoxes of power: indigenous peoples in the Permanent Forum. *Cooperation and Conflict* 46(4), pp. 543–562.

Lynge, A. (2002). *The Right to Return: Fifty Years of Struggle by Relocated Inughuit in Greenland. Complete with an English translation of Denmark's Eastern High Court Ruling*. Nuuk: Atuagkat.

Lynge, F. (1999). *Selvstændighed for Grønland?*. Copenhagen: Arctic Information.

Mortensen, R. R. (2007). Sport, eksport og selvstændighedskamp. *Idrætshistorisk Årbog* 23, pp. 123–135.

Nielsen, J. (2001). *Greenland's Geopolitical Reality and its Political-Economic Consequences*. Working Paper. Copenhagen: Danish Foreign Policy Institute.

Nybrandt, M. and Mikkelsen, T. (2016). *Drømme i tynd luft*. Copenhagen: Forlæns.

Okalik, P. and Motzfeldt, J. (2000). *Erklæring vedrørende samarbejde med Nunavuts regering og Grønlands Hjemmestyre*. Available at: http://dk.vintage.nanoq.gl/Emner/Landsstyre/Departementer/Landsstyreformandens%20Departement_2013/Udenrigsdirektoratet/Hvad_arbejder_vi_med/Arktisk_Samarbejde/~/media/nanoq/Udenrigsdirektoratet/samarbejde%20med%20Canada/erkl%C3%A6ring%20vedr%20%20samarbejde%20nunavut%20%20%20gr%C3%B8nland.ashx. [Accessed 29 February 2016].

Qvist, N. (2016). Sprogforsker: Engelsk kan oversvømme grønlandsk. *Sermitsiaq*. Available at: http://sermitsiaq.ag/sprogforsker-engelsk-kan-oversvoemme-groenlandsk. [Accessed 1 February 2017].

Petersen, H., ed., (2006). *Grønland i Verdenssamfundet. Udvikling og forandring af normer og praksis*. Nuuk: Atuagkat.

Ricœur, P. (1988[1985]). *Time and Narrative*. Vol. 3. Chicago, IL: University of Chicago Press.

Sejersen, F. (1999). At være grønlænder – hvem sætter grænserne?. In: J. Lorentzen, E. Jensen and H. Gulløv, eds., (1999): *Inuit, kultur og samfund – en grundbog i eskimologi*. Århus: Systime, pp. 126–131.

Shadian, J. (2010). From states to polities: Re-conceptualizing sovereignty through Inuit governance. European Journal of International Relations 16(3), pp. 485–510.

Skydsbjerg, H. (2002). Selvstændighed men... *Sermitsiaq*. p. 22.

Søbye, G. (2013). To be or not to be indigenous: Defining people and sovereignty in Greenland after self-government. In: K Langgård and K. Pedersen, eds., *Modernity and Heritage: How to Combine the Two in Inuit Societies*. Nuuk: Atuagkat, pp.187–205.

Sommer, K. (2012). Ane Hansen støtter KNAPK's oplysningskampagne. *Kalaallit Nunaata Radioa*. Available at: http://knr.gl/da/nyheder/ane-hansen-st%C3%B8tterknapk%C2%B4s-oplysningskampagne. [Accessed 2 June 2015].

Sørensen, B. (1991). Sigende tavshed. Køn og Etnicitet i Nuuk, Grønland. *Tidsskriftet Antropologi* 24, pp.41–58.

Sørensen, B. (1994). *Magt eller afmagt? Køn, følelser og vold i Grønland*. Copenhagen: Akademisk Forlag.

Sørensen, H. (2016). Olsvig ønsker magtopgør i Arktisk Råd. *Kalaallit Nunaata Radioa*. Available at: http://knr.gl/da/nyheder/olsvig-%C3%B8nsker-magtopg%C3%B8r-i-arktisk-r%C3%A5d. [Accessed 1 March 2016].

Strandsbjerg, J. (2014). Making sense of contemporary Greenland: Indigeneity, resources and sovereignty. In: R. Powell and K. Dodds, eds., *Polar Geopolitics? Knowledges, Resources and Legal Regimes*. London: Edward Elgar, pp.259–576.

Taagholt, J. (2002). Thule Air Base 50 år. *Tidsskriftet Grønland* 50(2–3), pp. 41–112.

Thorleifsen, D., ed., (2003). *De vestnordiske landes fælleshistorie – udvalg af indledende betragtninger over dele af den vestnordiske fælleshistorie*. Nuuk: Inussuk.

Ugebrevet A4 (2013). *Grønlænderne kigger mod Canada*. Available at: http://a4.media. avisen.dk/GetImage.ashx?imageid=28241&sizeid=255. [Accessed 3 March 2016].

Veirum, T. (2016). Vestnordisk Råd ønsker observatørstatus i Arktisk Råd. *Kalaallit Nunaata Radioa*. Available at: http://knr.gl/da/nyheder/vestnordisk-r%C3%A5d-%C3%B8nsker-observat%C3%B8rstatus-i-arktisk-r%C3%A5d. [Accessed 1 March 2016].

2 Independence through international affairs

How foreign relations shaped Greenlandic identity before 1979

Jens Heinrich

As Greenland moves steadily closer to independence, the government of Greenland has begun to carve out space for a separate Greenlandic foreign policy, even though the Self-Government Act from 2009 between Greenland and Denmark lists foreign relations as a Danish responsibility. For instance, Greenland has opened up representations in Washington, Copenhagen and Brussels and is planning to open further representative offices in key countries. Foreign relations seem to be a key part of the independence project, but it seems pertinent to ask how far back one can track this link between Greenlandic identity and foreign relations. In other words, are Greenlandic foreign relations an integral part of the independence project or a recent addendum?

This chapter examines how foreign matters played a role in shaping Greenlandic identity between the turn of the nineteenth century and 1979 by focusing on how Greenlandic institutions, such as the provincial councils, reacted to international events and Denmark's behaviour in the international sphere.[1] Initially the Greenlandic politicians focused on internal matters, in accordance with the official policy, leaving foreign affairs to the colonial administration in Copenhagen. Denmark's Greenland policy aimed at an ever-increasing Greenlandic participation and responsibility. The goal was to let the Greenlanders learn to manage their own affairs through the councils. Over time, development and increasing maturity would lead to increasing participation and self-determination. When the desired maturity and ability were reached, a revision of the ruling of Greenland could be invoked. This happened in 1920, after the Second World War, and in the 1970s. As a growing ability and political awareness developed in Greenland, Greenlandic politicians became interested in international affairs, and Greenland's foreign relations interacted with identity dynamics and paved way for increased independence. The growth of a Greenlandic nationalism was perhaps an unexpected part of the political development.

Though I aim to describe a development over almost eighty years in this chapter, I specifically focus on major events that together illustrate how Greenlandic elites changed their understanding of the outside world and Greenland's identity. I argue that the development occurred in three periods: the period before the Second World War, the period during and immediately after the war, and, finally, the time leading up to the introduction of home rule in 1979. These periods were central in the development of the political culture in Greenland, and show a still greater participation from Greenlandic politicians. Greenlandic politicians over time gained a still greater political understanding and foreign matters played an important part in this process.

1900–1940: Danish control – Greenlandic spectators

In the period before the Second World War, Greenlandic society was deliberately kept away from international affairs and representatives of the Greenlandic community were only informed of Danish decisions concerning Greenland's international status *ex post facto*. In this period, Greenland was an isolated and protected Danish colony (the island became an equal part of the Danish kingdom in 1953). The isolation and protection of the Greenlandic society emanated from the idea of Greenlandic society being fragile, and therefore contact with foreign nations would lead to deterioration or even annihilation of the society. For instance, a 1931 book on Greenlandic civics contained a passage by former Danish interior minister, C.N. Hauge, which stated: "the Greenlanders are unable to protect themselves. Denmark is able to do this and Denmark is obliged to do so" (Oldendow 1931, 15). On this account, the Danish colonial administration sought to strengthen the society and eventually Greenland could be opened up to the rest of the world. One of the means used was education of the Greenlanders through participation in administrative councils. The ultimate goal was to help the Greenlanders become capable of managing their own affairs (Boel and Thuesen 1993, 38; Heinrich 2012, 48). Initially a board of guardians (Forstanderskabsrådene) with limited responsibility was formed around 1860 and consisted of equal numbers of Greenlandic and European members. Each board covered a colonial district. In 1911 two provincial and 62 communal councils were established to substitute the former boards and the councils became the Greenlanders' main institutional voice vis-à-vis Denmark and the place where one would expect to find concerns about foreign affairs.

Danish paternalism can be illustrated by the lack of involvement of representatives of Greenlandic society in major international decisions and events that concerned the island. In this period, Denmark fought a diplomatic battle to gain international recognition of its sovereignty over Greenland, which was only confirmed by the outside world with the verdict of the International Court in The Hague in 1933. The Greenlandic councils did not take part in major foreign policy decisions in that period. In 1916 Denmark sold the Danish West Indies (now the US Virgin Islands) to the US. Part of the sale was an agreement in which the US authorities recognised the Danish right to all of Greenland. In a following correspondence between the US secretary of State Hughes and the Danish consul to Washington, the Monroe doctrine was underlined. Territories on the American hemisphere would not be allowed to be transferred to other European nations (Hughes 1921). The sale was not conveyed to the provincial councils, but the governors were informed that Denmark now had control of all of Greenland.

The provincial councils did not passively accept being left out of major decisions. For instance, in 1924 an agreement was made between Denmark and Norway concerning scientific activities and hunting in East Greenland. The dispute between Denmark and Norway concerned the uninhabited northeast Greenland. Denmark had established a colony, Scoresbysund, in 1924 to show the Danish flag and claim sovereignty. The coast north of the colony was uninhabited except for a number of hunters of Danish and Norwegian nationalities. The Danish authorities did not see their right to all of Greenland as being broken by the agreement, but it did support Danish–Norwegian relations. At a meeting of the provincial council in south Greenland later that year, the fact that the Danish government had made the agreement without consulting the provincial councils was criticised. The council felt their interests and the interests of the East Greenlanders could be compromised, and the cooperation between Danes and Greenlanders could suffer. In the eyes of the councils the governing law clearly stipulated that the councils should be included in such a matter (Government of Denmark 1925, 289–90).

The governor explained the circumstances of the agreement at a Northern Council meeting the same year (Government of Denmark 1925, 301–302). A reason for why the councils had not been included was not given, besides the fact that the areas in Greenland were not under the colonial districts and administered directly from Copenhagen and therefore not part of the matters pertaining to the councils. The Northern Council discussed the agreement at some length, and the lack of inclusion was criticised and it was implied that Denmark had shirked its responsibility as steward of the island's foreign policy. According to one council member, as "the Greenlanders weren't accustomed to deal with such [foreign] matters ..." (Government of Denmark 1925, 302) the Danes had an even greater responsibility to make sensible decisions and the agreement jeopardised the welfare of Greenlandic society and the cooperation between Denmark and Greenland.

Norway, despite the agreement with Denmark, saw parts of the northeastern coast as *terra nullius* and claimed these areas to be Norwegian. In 1931 Denmark filed a case at the International Court in The Hague, which found the Norwegian occupation to be unlawful, thus recognising Denmark's legal authority over all of Greenland (Ministry of Foreign Affairs 1933). The verdict pleased the provincial councils, and they expressed the hope of "a still more cordial connection between our country and Denmark and years of peaceful cooperation for the honour of Denmark and the benefit of Greenland" (Government of Denmark 1938, 83).

Prior to 1940 foreign affairs was considered a Danish responsibility, but the provincial councils did have an attitude towards these matters and expected to be included in the matters, or at least to be heard. Even though the provincial councils could express their opinion and should be heard in relevant matters according to the governing law, the Danish colonial administration was not obliged to follow their wishes. The colonial administration did not see a discrepancy between the idea of cooperation with the Greenlanders and the lack of inclusion on decision-making concerning foreign matters. The provincial councils did express faith in the colonial administration to handle the interests of Greenland, but at the same time Denmark did have interests of their own – and these two approaches were not always aligned.

1940–1951: Greenland joins the world

When Denmark was occupied by Germany in April 1940 Greenland was suddenly on its own. The above-mentioned 1925 law concerning the ruling of Greenland gave the two governors the right to rule the island on behalf of the Danish government. In the beginning of May 1940, a joint meeting of the two provincial councils was held. At this meeting the councils gave their support to the governors. This step was not a mere formality and the councils actually had some power. Within the Danish authorities, there was a fear that Greenland sought to get rid of the Danes and perhaps turn towards the US for support (Brun 1941). This meant that the governors had to respect and listen to the councils that otherwise only had advisory status. The war and the subsequent isolation from occupied Denmark gave the two governors the *de jure* authority over Greenland's foreign relations. The Greenlandic politicians at the time recognised their shortcomings and inexperience in this area and they accepted and relied on the governors' authority. The governors saw it as their primary objective to maintain Greenland as part of Denmark, and this required taking care of their Greenlandic subjects and thereby maintain the Greenlandic loyalty towards Denmark.

However, the power vacuum meant that other actors also played a crucial role in Greenland's foreign policy. The Danish consul in Washington, Henrik Kauffmann, had declared himself independent of the Danish government in Copenhagen and his position close to the decision-making circles in Washington meant that he *de facto* had a substantial influence over Greenland's foreign policy. The government of Denmark had maintained its work and had initiated a cooperation agreement with the occupying German forces. Kauffmann was convinced the government to be under German influence, and he followed an independent political course to further what he perceived to be Danish interests, most importantly to gain American recognition of Denmark as belonging to the Allied powers (Lidegaard 1996; Løkkegaard 1968). The United States had important interests in Greenland, which Kauffmann could use as leverage. Greenland had the only mine in the world producing cryolite. Cryolite was used in the production of aluminium, and therefore important in the war production (Heinrich 2012, 90–95). Kauffmann also used profits from the cryolite mine to bankroll his diplomatic activities (Lidegaard 1996). Another American interest in Greenland was its strategic geographical location right between America and Europe, and Greenland came to function as a step for airplanes travelling to Europe (Ammendrup 2007, 18).

Kauffmann had a clear interest in controlling Greenland and thereby letting its importance work to his advantage. He negotiated the 1941 defence agreement with the US authorities. Article 10 in the agreement was of especially controversial – it stated that the agreement should be in effect until the current threat towards the American continent had passed. The governors were frustrated from being excluded from the final negotiations and for being forced to sign the agreement with short notice. Kauffmann had warned the governors of an impending Canadian invasion if they did not sign the agreement. Both governors regarded the US as the most appropriate partner under the circumstances, and they fully recognised the need for the agreement, but objected to the way the agreement had been carried out. The government of Denmark did not recognise the agreement and filed a recall of Kauffmann. The signing of the agreement actually won Kauffmann a major victory, as he was officially acknowledged as representing Denmark by US authorities. Until this point Kauffmann needed the approval of the governors, as they were seen as legitimate representatives of the government of Denmark (Løkkegaard 1968, 89). The governors and government of Denmark saw the agreement and especially article 10 as a serious threat to Danish sovereignty, as it was unclear when the Americans would actually leave Greenland (Lidegaard 1996).

The US maintained its interest in Greenland, also after the defeat of Germany and the Axis powers in 1945. Following the liberation of Denmark, the intermediate Danish government approved the 1941 defence agreement, and Kauffmann was made minister without portfolio in recognition of his efforts. The Greenlandic members of the provincial councils were not included in the negotiations leading up to the 1941 agreement, neither were the governors even though they had a legal claim to rule Greenland under the circumstances. The war had thus strengthened the councils' position vis-à-vis the governors, but the governors only had partial control over Greenland's foreign policy.

Although the councils' influence remained limited, the Second World War did change how both Greenlanders and Danes viewed Greenland and its developmental possibilities. The American presence and entrepreneurship showed that large-scale endeavours were possible, even though the harsh environment in Greenland formed a formidable obstacle. Industrial fishery with modern harbours and fishing plants came to be the economic dream of the future Greenland.

Furthermore, Greenland was no longer at the outskirts of the world, as Greenland suddenly became important for the war effort (Claeson 1983).

The Allied presence in Greenland was not unproblematic and there were instances where it challenged Danish sovereignty and created tensions between the different parties. The agreement allowed establishment of a number of US military bases. The largest base at Narsarsuaq had around 6,000 troops at its peak; in comparison the Greenlandic population totalled around 20,000 and the largest colonial centre had around 700 inhabitants. Danish protection of the small Greenlandic communities in that area was severely challenged. Already in 1940 an American (and a Canadian) consul had gone to Nuuk to mediate between the Danish/Greenlandic and US and Canadian authorities. This foreign presence provoked a number of Danes in Nuuk as it could jeopardise the Greenlanders' faith in the Danes (Vibe 1942).

A significant aspect of the experiences gained during the war was the feeling of unity within the Greenlandic society and in many respects the war became a catalyst for general tendencies in society. Even though the councils only held three meetings during the war and the real authority was governor Brun (Heinrich 2012, 125-28), one gets the sense that Greenland was no longer isolated and that it actually had a strategic importance and relevance to the outside world (Jakobsen and Heinrich 2005). The 1925 law concerning the ruling of Greenland had opened the possibility of joint meetings between the two councils, but fifteen years had to pass before the first joint meeting took place in May 1940 as a result of the circumstances.

In that sense, the 1925 law had created the foundation for more national unity in Greenland, but it was international events that finally activated these dynamics. At the 1943 meeting, governor Brun encouraged the members to formulate their thoughts on the future political course of Greenland. The encouragement led to a thorough debate the following years. It seems that these events and experiences gave the Greenlandic politicians a sense of a future political direction and knowledge of being able to handle difficult situations. There was a sense that the pre-war system ought to be altered; the Greenlanders should have a greater degree of participation and Greenland should to a greater extent be ruled from within Greenland. The most prolific Greenlandic politician of the era, Augo Lynge (1899–1959), wanted to develop the Greenlandic society in close cooperation with Denmark. He wanted to use Danish culture and language as means to this process, and in the following period, the Danification of Greenland was regarded as the most efficient way to develop Greenland. At the same time foreign relations were thought by both the Danish administration and by Greenlandic politicians to be a Danish matter. Actual equality was difficult to implement. The economic resources and productivity of Greenland, and the lower educational level caused lower wages for Greenlanders and fewer Greenlanders in leading positions. Both Danes and Greenlanders accepted the Danification of Greenland, but eventually it gave way for a growing Greenlandic nationalism or – as it came to be called – Greenlandification.

However, though one can trace a nascent awareness of Greenland's role in the world, one should not exaggerate the impact of the war. Foreign relations remained dominated by Denmark and the councils largely refrained from opposing this state of affairs. For instance, the provincial councils did not show any interest in defence matters during the war. At the 1941 joint meeting of the councils, which took place in Nuuk in June, Governor Brun presented the defence agreement and told the members to embrace it, as it ensured "the position of Greenland, its cohesion with Denmark and the interests of Greenland" (Greenland National Council 1941, 6). The matter did not raise any debate or questions from the Greenlandic politicians. Similarly, in 1951 a new defence agreement was signed between the Danish and US governments, now both members of NATO (Danish Foreign Policy Institute 1997; Danish Institute for International Studies 2005; Lidegaard 1996, 480–501). The agreement was presented for the Greenlandic National Council at its 1951 meeting (Greenland National Council 1951, 49–50).[2] The agreement did not raise any debate or questions at the meeting, or any objection against the fact that Denmark had negotiated the agreement without including any Greenlandic representatives. In Thule a hunters' council had since the 1920s been the official voice of the community. According to the minutes from their 1950 and 1951 meetings, the defence agreement was not mentioned by either the Danish representatives or by the local members (Government of Denmark 1951).

1951–1979: A new movement for independence

Greenland experienced significant changes, partly prompted by international developments, in the post-war period. During the global wave of decolonisation that followed the war, it had become inconvenient for Denmark to maintain Greenland as a colony and Copenhagen therefore abolished Greenland's status as a colony and made it a county within Denmark in 1953. Besides the global pressure for decolonisation, the change of status also reflected a Danish political wish to continue the relationship, as Denmark saw it as its historical responsibility. Greenlandic politicians also wanted to develop Greenlandic society in close connection with a solid nation as Denmark (Heinrich 2012; Beukel et al. 2010).

A new and more nationalistic generation of Greenlandic politicians emerged in the beginning of the 1970s (Sørensen 1983, 225). Most of these had had their education in Denmark and had been members of the student organisation UGR (Young Greenlanders Councils) and came to represent a turn towards Greenlandic nationalism and involvement in foreign affairs that had been brewing since the 1950s. A top-down development programme for Greenland had begun in the 1950s and 1960s and had created new conditions and infrastructure, such as schools, hospitals, and apartment complexes. The underlining premise was that Greenland should become a welfare state that resembled Danish society. This was supported by leading Greenlandic politicians, but it also led to a growing dissatisfaction and gave rise to a new political agenda that stressed independence from Denmark. Amongst Greenlanders, the modernisation was seen as steps towards equality with Denmark. But in the 1960s, many Greenlanders still felt inferior to Denmark and critiques of the development program began to be voiced (Jensen 1996, 155–156). Almost all leading positions were held by Danes, and Greenlanders doing similar work had lower salaries (Greenland Commission of 1960, 149). The reason for the difference in salaries was that the Greenlanders holding similar jobs as the Danes were to be paid according to other Greenlandic professions (e.g. fishers and hunters), and at the same time Danes were given a bonus for working in Greenland. Greenlandic salaries were based on the productivity of the society (Thorleifsen 2003, 107). Many Greenlanders felt like being spectators to the development program, without having any say in the matter. This dissatisfaction affected several elections in the early 1970s. In the 1971 National Council election, members of the new generation of more nationalistically minded politicians, such as Jonathan Motzfeldt and Lars-Emil Johansen, were elected and later the same year Moses Olsen, another key figure of this movement, was elected to the Danish parliament. Moses Olsen announced in his first speech as a member of the Danish parliament that he aimed to alter the status of Greenland and to put an end to Denmark's long-distance management of Greenland (Olsen 1971). These politicians perceived the modernisation policy as an attempt at "Danification", aimed at obliterating Greenlandic culture and language and replacing it with Danish culture and language. They saw themselves as able to manage own affairs, but did not want to break the connection between Greenland and Denmark. They wanted a more Greenlandic Greenland, run by Greenlanders in cooperation with Denmark in actual equality.

International relations played an important role in this movement as became evident in the 1972 EEC membership referendum. Denmark had since 1961 worked to join the European Economic Community and in 1972 a referendum was held on whether or not to join the community. Even though 70 per cent of the votes in Greenland went to the "no" side, the total votes from Denmark and Greenland gave 63 per cent for a "yes". Greenland was part of Denmark and joined the community. The Faroe Islands, also a part of the Danish Kingdom, had had extensive autonomy (home rule) since 1948 and it was therefore able to avoid joining the community. Greenlandic leaders saw EEC membership as a threat to the Greenlandic economy, as joining the EEC would diminish Greenland's ability to exploit its marine resources, especially cod. Being a member would mean other member states would have free access to Greenlandic waters. Greenland had since 1953 had two members of the Danish parliament (Folketinget). These members had initially supported the process and the National Council also supported it, but it wanted certain special arrangements to protect Greenlandic interests, such as fisheries.

Other international concerns also exacerbated tensions between Denmark and Greenland and fuelled the push for increased autonomy. In the midst of the international oil crisis, Danish politicians saw a potential in Greenland as a means to relieve the energy crisis of the Western world. Explorations in Greenland had shown vast reserves of oil and other natural resources. Leading Greenlanders feared that the island would be exploited by Denmark and foreign companies. The right to the riches of Greenland's underground gave rise to political mobilisation in Greenland and it put a strain on the relationship between Greenland and Denmark (Sejersen 2014).

In 1972, a National Council committee proposed a commission to examine how a home rule arrangement could be implemented in Greenland (Greenland National Council 1972, 304). Eventually a Greenlandic–Danish commission was formed and in 1979 home rule was introduced, which meant a transfer of power from administrative and political bodies in Denmark to Greenland. And within the boundaries of the Home Rule Act, the Greenland Home Rule Government now had the right to legislate (Government of Denmark 1978). The newly formed government of Greenland took it as one of their first tasks to prepare a referendum on Greenlandic membership of the EEC. At the 1982 referendum, 52 per cent of voters wanted to leave the community, while 46 per cent wanted to stay and in 1985 Greenland thus became the first community to leave EEC. An agreement with the EEC was negotiated and it gave Greenland status as an "overseas country or territory" and thereby continued relations with Brussels. An important step towards independence had been taken and Greenland's foreign relations had been reoriented.

Conclusion

Foreign relations have had a great impact on the political development of Greenland. As Greenland and the Greenlanders began to see themselves as related to the outside world, the first ideas of a future independence began to emerge. However, it took quite a bit of time before this awareness of Greenland's potential role in the world was formulated as an actual foreign policy vision. Being able to take part in the rest of the world and having something to offer have had an impact on how Greenlandic society came into being.

The Second World War ended the isolation of the island, and the effects of the war meant a revision of how Greenland was ruled was initiated. The war also saw the start of the American presence in Greenland, and today Greenland is still a significant part of the US military infrastructure. The American presence continued during and after the Cold War. The Danish colonial policy aimed at maturing the Greenlandic population, and making Greenland able to manage its own affairs. From an early point the Greenlanders embraced this policy, and strived to gain still greater participation. The political mobilisation in the wake of the modernisation process meant a growing demand for participation, which resulted in the independence movement, which achieved home rule in 1979. The policy of promoting a still greater participation on own affairs have characterised the development of the Greenlandic society since the middle of the nineteenth century. This has been done in cooperation with the Danish government.

Along the way there has been a growing Greenlandic political awareness. For instance, a critique of the premises of the modernisation policy was formulated beginning in the 1960s. The younger generation of Greenlanders in this period opposed the idea of making Greenland into a Danish society. The modernisation process in their opinion needed to change. The Danish remote control of Greenland could not continue. Greenland should be ruled by Greenlanders and ruled from Greenland, which came to be the essence of the Greenland Home Rule Act. This also meant a growing Greenlandic participation regarding foreign matters, but still with certain crucial matters under Danish jurisdiction.

Notes

1   Greenland had two provincial councils from 1911–50 – one in the south and one in the north, covering the west coast. East Greenland and the Thule area were included in the 1960s. In 1951 the two councils were merged in to one. The councils had advisory status, but were usually included in relevant matters pertaining to Greenland. The Danish parliament (Rigsdagen till 1953, and hereafter Folketinget) was the legislative body. Greenland had two members of the Danish parliament from 1953.
2   In 1950 the two provincial councils were merged into one council covering all of the west coast of Greenland.

References

Ammendrup, H. (2007). *Grønlands Styrelse 1939–48: et direktorat i spidsen for udviklingen?* Copenhagen: University of Copenhagen.
Beukel, E., Jensen, F., and Rytter, J. (2010). *Phasing out the Colonial Status of Greenland, 1945–54,* Copenhagen: Museum Tusculanum Press.
Boel, J. and Thuesen, S. (1993). Grønland og den store verden: 2. verdenskrigs betydning for det dansk-grønlandske forhold. *Grønlandsk kultur- og samfundsforskning* 2, pp. 34-61.
Brun, E. (1941). Letter to Consul Kauffmann, May 17. National Archive of Denmark. Copenhagen.
Claeson, P. (1983). *Grønland Middelhavets perle: Et indblik i amerikansk atomkrigs forberedelse.* Copenhagen: Eirene.

Danish Foreign Policy Institute. (1997). *Grønland under den Kolde Krig, Dansk og Amerikansk Sikkerhedspolitik 1945–68*. Copenhagen: Danish Foreign Policy Institute.

Danish Institute for International Studies. (2005). *Danmark under den Kolde Krig, Den Sikkerhedspolitiske Situation 1945-1991*. Copenhagen: Danish Institure for International Studies.

Danish Ministry of Foreign Affairs. (1933). *Haag-Dommen af 5. April 1933 om Østgrønlands Retsstilling*. Copenhagen: Nyt Nordisk Forlag.

Government of Denmark. (1925). *Beretning & Kundgørelser vedrørende Kolonierne i Grønland*. Copenhagen: H.J. Schulz.

Government of Denmark. (1938). *Beretning & Kundgørelser vedrørende Kolonierne i Grønland*. Copenhagen: H.J. Schulz.

Government of Denmark. (1951). *Beretninger vedrørende Grønland*. Copenhagen: Ministry for Greenland.

Government of Denmark. (1978). Lov nr. 577 af 29. november 1978 om Grønlands. Available at: www.retsinformation.dk/Forms/R0710.aspx?id=168699 [Accessed 22 May 2017].

Greenland National Council. (1941). *Grønlands Landsråds Forhandlinger*. Nuuk: Greenland National Council.

Greenland National Council. (1951). *Grønlands Landsråds Forhandlinger*. Nuuk: Greenland National Council.

Greenland National Council. (1972). *Grønlands Landsråds Forhandlinger*. Nuuk: Greenland National Council.

Greenland Commission of 1960. (1964). *Betænkning fra Grønlandsudvalget af 1960*. Copenhagen: Ministry for Greenland.

Heinrich, J. (2012): *Eske Brun og det moderne Grønlands tilblivelse*. Nuuk: Inussuk.

Hughes, C. (1921). Letter to Constantin Brun, August 3. National Archive of Denmark. Copenhagen.

Jakobsen, A. and Heinrich, J. (2005). *Sorsunnersuaq kingulleq nunarpullu – Anden Verdenskrig og Grønland*. Nuuk: Greenland's National Museum and Archive.

Jensen, E. (1996). Udvikling, oplysning, kultur. En skitse til Hans Lynges politiske liv. *Tidsskriftet Grønland* 44(3), pp. 151–162.

Lidegaard, B. (1996). *I Kongens Navn : Henrik Kauffmann i Dansk Diplomati 1919–1958*. Copenhagen: Samleren.

Løkkegaard, F. (1968). *Det danske gesandtskab i Washington 1940-1942: Henrik Kauffmann som uafhængig dansk gesandt i USA 1940-1942 og hans politik vedrørende Grønland og de oplagte danske skibe i Amerika*. Copenhagen: Gyldendal.

Oldendow, Knud (1931): *Den Grønlandske Samfundslære*. Godthåb: Sydgrønlands bogtrykkeri.

Olsen, M. (1971): Speech to the Danish Parliament. *Folketingstidende*, 21 October.

Sejersen, F. (2014). *Efterforskning og udnyttelse af råstoffer i et historisk perspektiv*. Copenhagen: University of Copenhagen.

Sørensen, A. (1983). *Danmark-Grønland i det 20. århundrede – en historisk oversigt*. Copenhagen: Arnold Busck.

Thorleifsen, D. (2003): Kampen for etnisk identitet og krav om ekstern selvbestemmelsesret. In: D. Thorleifsen, ed., *De vestnordiske landes fælles historie*. Nuuk: Inussuk, pp. 105–115.

Vibe, C. (1942). Letter to Mikael Gam, June 25. National Archive of Denmark. Copenhagen.

3 Greenlandic sovereignty in practice

Uranium, independence, and foreign relations in Greenland between three logics of security

Kristian Søby Kristensen and
Jon Rahbek-Clemmensen

One of the key questions facing Greenland observers is how the island will orient itself vis-à-vis external actors – Denmark, foreign nations, companies, and NGOs among others – as it further develops its foreign relations. One thing seems evident from the literature: Greenland essentially strives to become a Westphalian nation-state and Greenlandic politics are driven by a yearning for sovereignty and independence, which is likely to shape how Nuuk faces the world (Gad 2014; Gerhardt 2011). However, when analysing Greenland's foreign relations, analysts need more concrete schema for understanding how Greenlandic policymakers approach specific issues. It is the meeting point between independence, sovereignty, and concrete matters that shapes how the island orients itself towards the world. Understanding how threats to future independence is translated into tangible politics in public discourse reveals both the limits and opportunities faced by outside actors when interacting with Greenland and the nature of the political setting in Greenland.

This chapter examines how Greenlandic policymakers debate concrete issues and how different dangers with different political logics threaten future independence and therefore shape how Greenland views its relationships to outside actors. It finds these political logics by examining the 2013 and 2014 debates on uranium extraction. Few issues have been more divisive than the question of whether and if indeed how uranium deposits should be extracted in Greenland. The question has fuelled heated debates, led to tensions between Nuuk and Copenhagen, and was a deciding factor in forming the 2014 coalition government (Nielsen 2014). The uranium debate was consequently not just a trivial matter of internal discussion, it was very much about Greenland, its future, and its independent role in the world – not least in relation to Denmark. The fiery discussion forced political actors to reveal different visions for Greenland's future and by unpacking how independence and identity are both threatened and secured in this political debate we can analyse the boundaries – the limits and opportunities – of the island's foreign relations.

The chapter uncovers three political logics of security and independence that establish the boundaries in Greenland's public debate – a political, an economic, and an environmental debate – and it argues that the debate of concrete issues is shaped by a Westphalian vision for Greenland as an independent and sovereign nation-state. We call these logics of security because political actors argue that Greenland's democratic system, its economic balance, and the sustainability of its environment are threatened in the uranium debate, and with these dangers follow threats to key topics necessary for genuine future Greenlandic independence. This reveals what needs to be secured for Greenland to be what it ought to be, and the core logics for debating Greenland's future. These political logics of security are intimately tied to questions of sovereignty and independence. Depending on whether uranium extraction is constituted as an issue of politics, economy or ecology, uranium extraction becomes both an existential threat to Greenland's future and a necessary step to secure this future. The three logics of security are also defined by specific silences, as the Greenlandic public shies away from discussing political, economic, and environmental issues that question the project of a sovereign and independent Greenland. All three logics are, therefore, largely Westphalian and they depict Greenland as a nascent nation-state.

The chapter consequently proceeds in three parts. First, we present an analytical framework for analysing Greenlandic political discourse drawing on the concepts of securitisation and sectors as developed by the so-called Copenhagen School. Next, we empirically analyse Greenlandic political discourse as it is manifested in the debate about uranium extraction and show how the three different political logics of security produce different threats as well as visions of an independent Greenlandic future. Finally, we use this exemplar of differences within Greenlandic political imagination to discuss how Greenland orients itself towards the world and the challenges and opportunities that face both Greenland and external actors as they interact with Greenland.

Analytical framework: politics as security and the logic of sectors

The so-called Copenhagen School offers a discourse analysis framework for distinguishing between the different logics that political actors use publicly (Buzan et al. 1998; Wæver 2002; Hansen 2006). The theory states that the nature of politics revolves around *security*: political actors garner support for their political goals by articulating a referent object as threatened and advocating that certain policies are enacted to ward off the threat. Issues are thought to be *politicised* if they are successfully framed as threatened, but the advocated policies are found amongst the normal political procedures of the state and *securitised* if extraordinary measures are invoked (Buzan et al. 1998, 23–26). Identifying different referent objects that are being framed as threatened and how different means are invoked to protect these objects allows us to understand how independence is linked to substantive issues and policies in Greenlandic political debate.

The concept of *sectors* enables analysts to disaggregate political discourse to investigate the substance of different political logics. We understand sectors to be made up of unique logics of security concerning a particular referent object defined in a particular and specific way and articulated as such by relevant political actors (this definition builds upon the one offered by Buzan et al. 1998, 27–29). For example, environmental security constitutes a separate sector only because political actors articulate the environment as threatened in a specific way and they suggest specific policies aimed at diminishing this specific threat. Sectors are thus empirical observations that analysts identify through discourse analysis and there is no *a priori* reason for the existence of specific sectors (Albert and Buzan 2011, 415–22). Though Buzan, Wæver, and de Wilde (1998, 7) argue for five sectors (the military, political, economic, societal, and environmental sectors), our application of their framework identifies three specific sectors in the Greenlandic political discourse: the political sector, the economic sector, and the environmental sector. All three display distinct political logics, threats, and distinctly conceived referent objects.

The political sector is defined by internal and external security threats against the state, political order, and democratic decision-making procedures (Buzan et al. 1998, 142–43). States aim to provide security and services for their citizens (output-legitimacy) while ensuring that their decision-making procedures are defined by legal or processual consistency (input-legitimacy). Besides the obvious threat from rebel groups or criminal gangs (none of which are relevant for the Greenlandic case), the main domestic threat potentially exists in the political arena, where both government and opposition politics can be framed as a threat to the effectiveness or legitimacy of the state. External threats include other states who might threaten the territorial integrity or the political sovereignty of the state or transnational phenomena, such as international organisations, multinational companies or even globalisation, which may erode the state's ability to exert effective and legitimate control over its territory and limit its ability to independently pursue its political goals.

Economic security is defined by threats against economic actors or institutions, including firms, socio-economic groups, the market, or the state's revenue streams. States can respond to foreign economic threats through foreign and domestic policy. The former includes manipulating the global terms of trade by erecting trade barriers, or through state intervention, although these policies are difficult to utilise by small actors such as Greenland (Buzan et al. 1998, 95–117). The latter category, which is also applicable to domestic economic threats, focuses on affecting the market through fiscal, industrial, or business policy.

Environmental security addresses threats against the environment. Buzan, Wæver, and de Wilde (1998, 71–94) highlight that analysing environmental security discourses entail examining whether political actors and audiences are aware of environmental problems, whether they accept responsibility for handling them, and how they imagine these problems can be solved. It is also crucial if an issue is articulated as a local, regional, or a global problem (Buzan et al. 1998, 84–91). Furthermore, the environment is rarely valued only for its own sake, but is rather linked to other referent objects, including identity, economics, and individual health; analysing how these links are made enables analysts to understand how the issue is being framed.

Buzan, Wæver, and de Wilde view identity as a potential referent object and societal security therefore constitutes a separate sector. However, the uranium debate reveals that identity issues are rarely articulated explicitly as being threatened. Discussions about identity issues are not absent from Greenlandic political discourse, but identity operates more through its ability to bestow meaning on other political discussions. Debates about political procedures, the economy, or the environment are inadvertently also discussions about what it means to be a Greenlander living in Greenland and although these debates are rarely taken in the open, Greenlandishness is constructed within debates about other issues. In the uranium debate, identity issues constantly materialise as references to Greenlandic independence. The discussion, accordingly, was not just about uranium, but about what uranium means in relation to what it means to be a Greenlander and how Greenland can become an independent and sovereign entity.

In sum, the Copenhagen School provides an analytical framework for analysing and understanding Greenlandic politics as debates about security as they relate to Greenlandic outside relations and as they are about different visions about how to realise a potentially endangered future independence. Departing from Buzan, Wæver, and de Wilde's sector approach and the emphasis on different kinds of security logics allows us to identify three different logics present in the Greenlandic debate. It further gives us questions that can be applied for each of these sectors, including whether issues are securitised or politicised, how threats and threatened objects are constructed, how political strategies to secure these are argued and justified and how the separate logics that characterise each sector play out. This provides us with answers to the identity struggle going on as Greenland debate and define its strategies for securing what is really at play: the future and survivability of a genuine Greenlandic political community. A community that is necessarily threatened as well as both constituted on and enabled by its specific relations to the outside world.

Three logics of security and the debate on uranium

Geologists have known about substantial uranium deposits in Greenland since the middle of the twentieth century, but substantial mining never took off and a ban on uranium mining ("the zero tolerance policy") was in effect from the late 1980s onwards. The ban meant that uranium mining, as well as other mining projects with uranium as a side-product – most prominently the rare earths deposit in Kvanefjeld in southern Greenland – were non-starters. The ban thus blocked a potential source of income for Greenland. Even as Greenland acquired full autonomous authority over its natural resources with the 2009 Self-Government Act, the incumbent Inuit Ataqatigiit (IA) government was hesitant to allow uranium mining. Nevertheless, it did permit exploratory mining-activities in Kvanefjeld (Boersma and Foley 2014, 13). The Siumut-led government that came to power in 2013 wanted to abandon the zero tolerance policy altogether, but met fierce resistance from the major opposition parties IA (who wanted to keep the policy in place) and the Democrats (who wanted to abolish the policy but was dissatisfied with the lack of democratic debate about the issue) and from NGOs and the main newspapers. In the fall of 2013, the government parties pushed the policy through parliament with a one-vote majority. This caused widespread controversy, including the largest public demonstrations in Greenlandic history, and made one of the coalition parties as well as Siumut MPs leave the government. The issue remained controversial and became one of the main rallying points during the 2014 election campaign.

The following examines how the issue of uranium extraction was discussed in the time period around the 2013 vote and the 2014 election campaign. It argues that three political logics can be uncovered in the debate. Each of these entails conceptions of security as defined by the Copenhagen School, as they all articulate threats and discuss the extraordinary character of measures taken in relation to uranium extraction. The final referent object of these debates is, however, always Greenland's political future. Each of these will be unpacked in the following.

Uranium and political threats

Political logics of security were used widely in the Greenlandic uranium debate, as the integrity of Greenland's political institutions and its relationship to other states and actors were portrayed as threatened by both sides in the debate. However, the main focal point of the debate was not the lack of administrative capacity of Greenlandic authorities, but the legal and political legitimacy of the decision-making procedures through which the uranium decision is made. The uranium debate became a discussion about whether the government parties were circumventing democratic rules and procedures. Opponents of uranium mining challenged the legitimacy of the process. They argued that the information provided by the government parties was incomplete, too complex, and at times erroneous, that an expert report about uranium mining was published only a few days before the vote in parliament, that the decision was made by the slimmest majority possible, and that the law had not passed through the right parliamentary committees (e.g. Duus 2013a; 2013d; 2013e; 2013f; Mølgaard 2013c; 2013d; Hansen 2013a). Some actors even argued that the decision violated the special right to inclusion assigned to native populations under international conventions (*Arbejderen* 2013; Lund 2013b). The centre-right Democrats found themselves in an odd middle position. Originally proponents of uranium mining, they stressed that they opposed the *process* through which the law was passed. During the 2014 election, the party was still critical of the process that led to the 2013 decision, but the party emphasised economic arguments (see below) that led it to abandon its opposition to the law (Duus 2013b; 2014a; Hannestad 2014a).

The pro-mining supporters also made use of processual arguments by emphasising that many procedures had been put in place and that reports had been published and they on their side denounced the "opposition's twisting of facts" (Mair 2014; see also Duus 2013g; Mølgaard 2013b; Petersen 2013). These processual arguments became particularly prominent during the 2014 election campaign, when IA leader Sara Olsvig, in an interview to *Politiken*, a major

Danish newspaper, sent mixed signals regarding the uranium referendum that IA had proposed. Even when pressed by the reporter, Olsvig refused to state whether a future IA government would accept the referendum result no matter the outcome, which made it unclear if the referendum would actually hold any democratic decision-making value (Klarskov 2014a; 2014b). Pro-mining advocates accused Olsvig of making a laughing stock of both the general public and international investors. By "speaking with two tongues", being "anti-democratic", and "hurting democracy" (Hannestad 2014b; Duus 2014b; 2014c; 2014d) IA was, according to its opponents, setting aside democratic rules on the issue of uranium. Siumut, in a similar vein, argued that the existing zero-tolerance policy had been put in place by an unelected committee and that it consequently also lacked political and democratic legitimacy (Kielsen 2013).

The recognition of Greenlandic authority played a peculiar role throughout the debate. Concurrent with the heated debate about legitimate democratic decision-making procedures in Nuuk, the Danish government in Copenhagen argued that uranium was a foreign policy and security policy issue, which according to the 2009 Self-Government Act would give Copenhagen final authority over the matter. The Greenlandic government parties retorted that uranium mining was a resource policy issue and thus well within Nuuk's purview. A lid was put on this essentially constitutional issue by establishing a joint working group that removed the always contentious issue of Danish/Greenlandic relations from the centre of the nuclear debate (Vestergaard 2014, 61–63; Vestergaard and Thomasen 2014). Still, parts of the Greenlandic opposition saw this as an opportunity for blocking the proposal. When Danish prime minister Helle Thorning-Schmidt made a statement that implied that Denmark had final say over Greenlandic uranium policy, Johan Lund Olsen, the IA member of the Danish parliament, conceded that Denmark could possibly block uranium mining in Greenland (Thorning-Schmidt 2013; Duus 2013c; Thorsen and Djørup 2013; Thorsen 2013; Rottbøll 2013; *Sermitsiaq* 2013). This placed IA in a strange position, where the party simultaneously argued for a Greenlandic referendum, sent mixed signals whether the IA government in Nuuk would follow the result of the plebiscite, while also arguing that actual Greenlandic sovereign authority over the issue was in question.

The uranium issue also highlighted that Greenland had responsibilities towards the international community. All actors – pro-mining and anti-mining – argued that Greenland had a special duty towards the world community to ensure that the mined uranium did not fall into the wrong hands. The opposition argued that the government parties had not established an overview of the plethora of international agreements and safeguards needed, while pro-mining voices argued that this had already being done (Petersen 2013). Uranium mining thus offered both a risk and an opportunity for Greenland by giving it a chance to demonstrate its maturity to the world community. Underneath this debate was a shared recognition that international obligations and the recognition of other states matter. Establishing safeguards and adhering to international treaty obligations was thus a way of demonstrating maturity to the international community and thus moving closer to international recognition.

While much ink was spilled over the legitimacy of the political decision-making process, the administrative capacities of the government of Greenland were rarely questioned in the public debate. Uranium management experts have since questioned whether the Greenlandic administration was ready to manage uranium mining according to international safety standards. For instance, recent historical studies show that the zero-tolerance policy was not based on a formal decision, which indicates that the institutional memory and administrative capacities of the Greenlandic resource agencies were rather weak (Vestergaard and Thomasen 2015; Thomasen 2014). The silence surrounding this dimension of the debate seems to be linked to the question of sovereignty and independence. Greenlandic political voices would struggle to question the maturity of the Greenlandic state, as such a discursive move would contradict the independence project and thus the central narrative in public discourse.

The uranium debate thus activated a discussion about the political legitimacy of the government of Greenland, both internally vis-à-vis its own citizenry and externally in its relationship to Copenhagen and the international community writ large. In a postcolonial polity, constantly in need of justifying its own political maturity to itself, to its interlocutors within the Danish realm, and to a global audience political legitimacy is no small matter. For instance, the referendum debate became a central and contentious issue. What from one side is seen as an extraordinary measure securing the legitimacy of the Greenlandic political system, can equally be seen as an indication of immaturity and thus a threat to the very same political system.

Uranium and economic threats

Greenland's economic well-being was often framed as being under threat, especially by pro-mining advocates, throughout the uranium debate. Greenland suffers from a significant structural deficit, projected to reach 10 per cent of GDP by 2030 (Economic Council of Greenland 2013). Following a liberal economic logic, both sides of the uranium debate agreed that creating opportunities for foreign investors was an absolute necessity for Greenland's future economic viability. The main point of contestation was *how* Greenland could create a fertile environment for investors. Pro-mining voices claimed that the state's fiscal and employment needs made uranium mining necessary (Nielsen 2014; Kruse 2014c; 2014b; Petersen 2013). This was countered by experts who criticised the government parties' economic estimates for being too optimistic and who argued that the possible income streams from uranium mining were too small to plug the hole in the government finances (Rosing 2014).

The opposition focused its arguments on political and environmental logics of security and, when addressing the economic side of the issue, it emphasised that Greenland should focus on other investment opportunities. It also mentioned that the economic consequences of uranium mining had yet to be explored and it emphasised the negative impact that it might have on Greenlandic agriculture and traditional occupations, such as hunting and fishing (e.g. Duus 2013e; Mølgaard

2013d; Lund 2013a; Kleist 2014; Olsen 2013). In other words, the opposition challenged the government parties' calculation of opportunity costs to the Greenlandic economy and the economic risks associated with uranium extraction. The government parties retorted that uranium mining would not interfere with other more traditional occupations (Kielsen 2013).

As Buzan, Wæver, and de Wilde point out, it is typically difficult to securitise the economy per se. Companies are meant to go bankrupt from time to time. Instead, economic security is often linked to other cherished objects that become threatened due to economic issues or problems (Buzan et al. 1998, 100). In Greenland, economic security is a precondition for *independence*. For instance, in a controversial speech given to an international audience in Iceland in October 2013, Aleqa Hammond, the then-premier, claimed that Greenland could become "a significant uranium exporter – among the world's top 10 or possibly top 5", which would be part of "the necessary transformation of the Greenlandic economy towards mining and oil and gas-related activities", one of "the necessary steps, which will enable Greenland to achieve independence one day within my lifetime." (Hammond 2013). This striving for independence significantly raises the stakes involved in the uranium debate. Independence became a crucial international identity marker that distinguished Greenland from the rest of the Inuit community, by framing it as a significant first amongst equals or as an Inuit trailblazer in the struggle for political independence. In Hammond's words, "Greenland is therefore today in the unique position of being the only indigenous people in the Arctic, which has its own government that has a recognised and agreed right to independence" (Hammond 2013).

The close link between Greenland's economic woes, Greenlandic independence, and its position as champion for Inuit self-assertion made the economic challenges particularly important. During the 2014 election campaign, pro-mining voices could consequently use economic arguments to castigate the uranium referendum that IA promised to hold if elected (Meinecke 2014; Klarskov 2014a; 2014b; Hannestad 2014b). The very suggestion of having a referendum, critics argued, would scare away investors, understood of course, within the subtext that this would jeopardise the economic foundation of a Greenlandic political future. As one mining executive put it,

> [i]nvestors don't know whether their projects would be stopped all of a sudden whenever there was a new political majority. ... All of the money they put into preparation would be gone. Worthless. That sort of thing has repercussions for the financial markets. They would find other projects to place their money in, should any doubt arise about Greenland's prospects.
>
> (Lindkvist 2014)

A referendum and the majority's ability to determine policy, both legitimate political procedures, were portrayed as politically immature and thus threats against Greenland's economic future and it could be argued that the rejection of these matters due to economic security concerns constitutes a securitising move.

The predictability needed for economic exchange and investment was argued as taking precedent over democratic authority and procedure. Appeals to economic security thus gave the government parties a trump vis-à-vis the opposition.

However, appealing to economic consciousness was a double-edged sword for the government parties. It could articulate its policies as necessities as long as its economic woes remained manageable. If the crisis slipped out of its hands, emergency loans or an increase of the block grant from Denmark would almost certainly lead to Danish calls for a renegotiation of the constitutional arrangement and thus a challenge to the dream of independence. Pro-mining advocates consequently refrained from pushing the logic of security too hard and they instead framed the issue as one of economic opportunities and employment without mentioning the possibility of fiscal break-down (Hannestad 2014b). The spectre of financial collapse was only mentioned occasionally and it was only elaborated in media that address non-Greenlandic audiences (Lanteigne 2014; Hansen 2013b; Gaardmand 2014). It illustrates the limitations that the current constitutional arrangements and Greenland's economic dependence on Denmark put on Greenland's public discourse. Pushing the logic of economic threats too far would only disrupt the sovereignty narrative by highlighting Greenland's lack of fiscal independence.

Uranium and ecological threats

Opposition voices used the potential ecological consequences of uranium mining as a strong argument against the government parties' position, justifying the need for a referendum. In that sense, environmental concerns were used in a securitising move to justify the need for extraordinary political procedures (a referendum) that trumped the normal majority-decision-making procedures in the Greenlandic parliament.

Mining critics vexed between arguing for an outright continuation of the ban to a call for a delay of the question to allow for more data collection and education (which meant keeping the door open for allowing uranium mining in the future) (e.g. Kruse 2014a). The latter part of the argument thus linked environmental concerns with political dissatisfaction regarding decision-making procedures. As IA's Kuupik Kleist, the former premier, put it,

> We are not ready to make this decision. The discussion unfolds as if uranium was something harmless that you could eat for dinner without problems. It is some dangerous stuff that can cause unpredictable environmental problems for a thousand years.
>
> (Jensen 2013)

Echoing the economic dimension of the debate, the opposition emphasised the *uncertainty* and *risk* involved in uranium mining to argue for a more ecologically precautionary approach (e.g. Duus 2013e; Lund 2013b).

Buzan, Wæver, and de Wilde argue that protecting the environment is often linked with other security logics and the Greenlandic case is no exception (Buzan et al. 1998, 75–79). According to its detractors, uranium mining would threaten not just the environment as such, but also the health of the Greenlanders and certain occupations, such as hunting and fishing (Kruse 2014c; Egede 2013; Kleist 2014). The latter were particularly important, because these activities were seen as crucial for the Greenlandic way of life. Hunting, fishing, and Greenland's peculiar nature made life in the north, and the people who inhabit the region, different from life in the developed south and thus made Greenlandic identity unique (Gad 2012, 218–19; 2014, 101; Jacobsen 2015). These arguments were largely implicit in the uranium debate, but they colour the arguments of each side by making uranium mining a threat to the essence of Greenlandic identity. Implicit in this line of argument was also a tension with the economic logic and a dispute about what kind of independence Greenland strived for. If economic development was the precondition for independence, but only at the price of giving up Greenland's unique way of life and close relationship to nature, then perhaps that kind of independence was not worth pursuing.

Pro-mining voices tried to refute these claims with three counter-arguments. First, some emphasised that the ecological and health dangers posed by uranium mining were exaggerated and they highlighted that the mining companies "comply with the highest possible standards of environmental protection" (Mair 2014; see also Kruse 2014b; Petersen 2013). Second, others argued that risks are impossible to understand for politicians and laymen and that it was important to trust experts (Frederiksen 2013; Mølgaard 2013a). This was effectively an attempt at depoliticising the question by making it a technical and/or administrative issue. Finally, when addressing an international audience, Aleqa Hammond tried to reorient the environmental discussion by highlighting that "nuclear power is one of the real mitigation options available in dealing with climate change today" (Hammond 2013). She thus implied that uranium mining helped tackle a much greater environmental problem (global warming), which could eclipse any local negative consequences simultaneously framing Greenland as a responsible member of international society.

Aside from Aleqa Hammond's statement, it was remarkable that both proponents and opponents of uranium mining cast the Arctic environment as a wholly *local* issue. Other voices in the High North articulate the Arctic environment as a *global* heritage that concerns citizens from the entire globe (Steinberg et al. 2014, 140–59). One could have imagined that Greenlandic actors would have picked up that discourse to frame themselves as having a special responsibility to *steward* the Arctic environment on behalf of international society. However, besides a few references to global warming, the debate cast the environment as having a local impact for the people of Greenland and their distinct identity. Greenlandic politics is about creating sovereignty over the territory to use it for its own purposes – such as hunting, fishing, or the energy industry. These are all potentially threatened by global environmental and climate politics – whether it is through NGOs, like Greenpeace, campaigning against seal hunting or the UN COP21 summit attempting to regulate greenhouse gas emissions.

Greenland and international recognition

Greenlandic identity and its striving for independence are ever-present. Even in debates on more pressing and practical threats and challenges involved in a practice as mundane as uranium mining. The threats facing Greenland are never just about the rules of Greenlandic politics, economic sustainability, or environmental uniqueness – they are about imagining and ultimately finding a road to independence that supports Greenland as a nation, defined by specific culture, language, and practices.

Greenlandic identity is caught between aboriginality and modernity, between a specific culture (and practices) and modern, Western concepts like democracy, welfare and market economy. Independence ties these seemingly contradictory notions together by creating a political horizon that has not been reached yet, but is always argued as reachable at some future point (Gad 2012, 219). As the uranium debate shows, independence turns politics into debates about security by providing a cherished object that is threatened, which, depending on definition and political contestation, bestows extraordinary value to political legitimacy, fiscal responsibility, and environmental pristineness.

These three logics simultaneously give substance to independence by providing a vision for a future Greenland and a logic for governing it until that vision is accomplished. Independence only becomes meaningful when it is made concrete, as a vision of an independent Greenland and its interaction with the outside world, defined by legitimate democracy, sustainable welfare, and a pristine environment. Each of them represents desires that Greenlandic politicians have to satisfy, which work to simultaneously enable and constrain both Greenlandic and external political actors.

Politically, Greenland equally needs outside recognition. The uranium debate shows that Greenland strives to become accepted as a fully-fledged member of international society and adhering to international standards is a key concern for Greenlandic politics. Observers concerned that an independent Greenland would be a reckless actor may find solace in the fact that Nuuk *wants* to adhere to international standards on issues of regional and international importance, such as uranium management, environmental protection, and fisheries management.

A key Greenlandic challenge in the coming years will be to go beyond official statements and ensuring that the Greenlandic state has the capacities needed to actually implement promised policies. In relation to the viability of Greenland's independence project it is troubling that the uranium debate all but ignored the Greenlandic state's *ability* to fulfil international standards, as this silence indicates that Greenlandic state capacity may be taboo in the public sphere.

Economically, Greenland continues to search for capital to close the gap in its finances. The uranium debate shows that Greenlandic politicians from all sides recognise the need for keeping Greenland open and attractive for foreign investors (although they disagreed about whether uranium mining was an essential industry). Foreign actors interested in interacting with Greenland must show that they somehow offer financial opportunities. Environmental NGOs, for instance, have to devise realistic plans for how a green Greenland can still become fiscally sustainable, maintain a high standard of living, and move towards independence. Environmental visions that *de facto* turn Greenland into a nature reservation are unlikely to be accepted by either Greenlandic public or political elites.

The uranium debate also showed that Greenlandic politicians prefer to avoid discussing the possibility and consequences of financial collapse. Economic sustainability was cast as a threat, but it was never a threat that would limit Greenland's future sovereignty. Barring a sudden resource boom, the long-term viability of the economy will most likely remain a key issue in the coming years, but outside actors (perhaps most importantly Denmark) will struggle to discuss these issues with Greenlandic leaders who prefer to avoid issues that question Greenlandic sovereignty.

Environmental appeals and arguments play a key role in Greenlandic public discourse. Independence is partly about preserving a specific way of life in nature, where hunting and fishing play essential, if often romanticised, roles and independence to some extent loses its meaning, if it is reached at the expense of the idea of a specific Greenlandic relationship to nature.

However, Greenlandic environmental debates are likely to be situated within a Westphalian understanding of nature that structures how international actors can make environmental arguments. Nature is a space owned by the Greenlandic people and an arena for the activities (like hunting and fishing) that defines Greenlandic identity. Appeals to Greenlandic stewardship on behalf of the international community were well-nigh absent in the debate on uranium. Environmental NGOs must decide if they want to challenge this conception or if they prefer to play along with it and focus their arguments on how environmental degradation hurt the Greenlandic way of life.

The three logics of security thus give an outline of the constraints and opportunities that will structure the relationship between Greenland and international actors. Of course, skilled political actors do not stay within existing confines produced by the salience of independence, but combine political logics to break existing conceptualisations apart. For instance, Greenland's political need for international recognition can be used to reconfigure the Greenlandic environmental discourse. But only if international actors show that environmental stewardship on behalf of the international community yields recognition or economic benefit. However, most basically the endangered object to be secured in Greenlandic political discourse is independence, and recognition as a genuine political community with legitimacy, authority and eventually sovereignty. The uranium debate showed that independence bestows meaning to all other discourses.

Conclusion and perspective

Our analysis of the uranium debate showed that it was about more than just uranium. It was basically about what kind of political community Greenland should strive to be – a debate that is necessary and important in any society and perhaps even more so in a community edging itself towards independence.

Greenlandic political actors evoke three overlapping visions and corresponding logics of governance – a political logic, an economic logic, and an environmental logic – that all play different roles in relation to independence, but simultaneously gain their political salience and power only through the key political objective of independence in Greenlandic politics. These logics, and perhaps precisely therefore, mostly come into play in a very Westphalian fashion, where Greenlandic policymakers strive to represent the Greenlandic state as politically competent, as economically sustainable, and as responsible for its own environment. At the same time, however, the island rests on a dual tension between modernity and aboriginality and between the local political community and the power of the international to both enable and constrain the Greenlandic strive for independence. This dual tension both structure and disrupts Greenlandic political discourse, when the power of foreign actors or Greenland's dependence on the outside world becomes visible.

Nevertheless, a key aspect of sovereignty is international recognition. For Greenland, sovereignty in practice, as is the title of this chapter, is about the international, and using the opportunities, relations and networks – economic, political, or environmental – that tie Greenland together with first and foremost arctic politics to substantiate and further Greenland's experiment with independence. For that, foreign relations and the international recognition that they provide are key, strategically, economically, and in terms of identity.

References

Albert, M. and Buzan, B. (2011). Securitization, sectors and functional differentiation. *Security Dialogue* 42(4–5), pp. 413–425.

Arbejderen. (2013): Største demo i Grønland i 29 år. Available at: http://arbejderen.dk/udland/st%C3%B8rste-demo-i-gr%C3%B8nland-i-29-%C3%A5r. [Accessed 4 September 2015].

Boersma, T. and Foley, K. (2014). *The Greenland Gold Rush: Promise and Pitfalls of Greenland's Energy and Mineral Resources.* Washington, DC: Brookings Institution.

Buzan, B., Wæver, O., and de Wilde, J.(1998). *Security: A New Framework for Analysis.* Boulder, CO: Lynne Rienner Publishers.

Duus, S. (2013a). Aviser: Der følger et enormt ansvar med. *Sermitsiaq.* Available at: http://sermitsiaq.ag/aviser-foelger-enormt-ansvar. [Accessed 4 September 2015].

Duus, S. (2013b). D ønsker både ja- og nej-sigere til demonstration. *Sermitsiaq.* October 23, 2013, Available at: http://sermitsiaq.ag/d-oensker-baade-ja-nej-sigere-demonstration. [Accessed 4 September 2015].

Duus, S. (2013c). Grønland og Danmark strides stadig om uran. *Sermitsiaq.* Available at: http://sermitsiaq.ag/groenland-danmark-strides-stadig-uran; [Accessed 4 September 2015].

Duus, S. (2013d). IA og D i fælles uran-opråb på falderebet. *Sermitsiaq.* Available at: http://sermitsiaq.ag/ia-d-i-faelles-uran-opraab-falderebet. [Accessed 4 September 2015].

Duus, S. (2013e). IA: Roligt nu, Per Berthelsen. *Sermitsiaq.* Available at: http://sermitsiaq.ag/ia-roligt-per-berthelsen; [Accessed 4 September 2015].

Duus, S. (2013f). Modstandere kræver folkeafstemning om uran. *Sermitsiaq.* Available at: http://sermitsiaq.ag/modstandere-kraever-folkeafstemning-uran; [Accessed 4 September 2015].

Duus, S. (2013g). SIS: Vi åbner ikke op for uranbrydning. *Sermitsiaq.* Available at: http://sermitsiaq.ag/sis-aabner-ikke-uranbrydning. [Accessed 4 September 2015].

Duus, S. (2014a). D-formand beklager uran-bommert. *Sermitsiaq.* Available at: http://sermitsiaq.ag/d-formand-beklager-uran-bommert. [Accessed 4 September 2015].

Duus, S. (2014b). Kritikere lugter blod efter Olsvig-forvirring om uran. *Sermitsiaq.* Available at: http://sermitsiaq.ag/kritikere-lugter-blod-olsvig-forvirring-uran. [Accessed 4 September 2015].

Duus, S. (2014c). Olsvig forklarer sig efter Politiken-artikel om uran og folkeafstemning. *Sermitsiaq.* Available at: http://sermitsiaq.ag/olsvig-forklarer-politiken-artikel-uran-folkeafstemning. [Accessed 4 September 2015].

Duus, S. (2014d). Trods massiv kritik: Olsvig fortryder ikke uran-udtalelser *Sermitsiaq.* Available at: http://sermitsiaq.ag/trods-massiv-kritik-olsvig-fortryder-ikke-uran-udtalelser. [Accessed 4 September 2015].

Economic Council of Greenland. (2013). *The Economy of Greenland 2013.* Nuuk: Government of Greenland.

Egede, A. (2013). Indlæg fra Inuit Ataqatigiit. Inatsisartut's item 88-1. Available at: http://www.inatsisartut.gl/dvd/EM2013/pdf/media/1136745/pkt88_em2013_nultolerancen_inuit_ataqatigiit_ordf_1beh_dk.pdf. [Accessed 4 September 2015].

Frederiksen, J. (2013). Indlæg fra Demokraatit. Inatsisartut's item 106-2. Available at: http://inatsisartut.gl/dvd/EM2013/pdf/media/1151871/pkt106_em2013_nultolerance_uran_ordf_2beh_demokraterne_dk.pdf. [Accessed 4 September 2015].

Gaardmand, N. (2014). Ingen ved rigtigt, hvad Grønland skal leve af. *Jyllandsposten Finans.* Available at: http://finans.dk/protected/finans/politik/ECE7236356/Ingen-ved-rigtigt-hvad-Gr%C3%B8nland-skal-leve-af/?ctxref=ext. [Accessed 4 September 2015].

Gad, U. (2012). Greenland projecting sovereignty: Denmark protecting sovereignty Away. In: R. Adler-Nissen and U. Gad, ed., *European Integration and Postcolonial Sovereignty Games: The EU Overseas Countries and Territories.* Abingdon: Routledge, pp. 217–34.

Gad, U. (2014). Greenland: A post-Danish sovereign nation state in the making. *Cooperation and Conflict* 49(1), pp. 98–118.

Gerhardt, H. (2011). The Inuit and Sovereignty: The Case of the Inuit Circumpolar Conference and Greenland. *Tidsskriftet Politik* 14(1), pp. 6–14.

Hammond, A. (2013). Tale ved Arctic Circle konference. Speech given at the 2013 Arctic Circle conference. Available at: http://naalakkersuisut.gl/~/media/Nanoq/Files/Pressemeddelelser/ARCTIC%20CIRCLE%20presentation%20TEKST%20DK.pdf. [Accessed 4 September 2015].

Hannestad, A. (2014a). Nyt opgør venter om Grønlands uran. *Politiken.* Available at: http://politiken.dk/oekonomi/dkoekonomi/ECE2414706/nyt-opgoer-venter-om-groenlands-uran/. [Accessed 4 September 2015].

Hannestad, A. (2014b). Uran fortrænger fisk i den grønlandske valgkamp. *Politiken* (27 November 2014), p. 3

Hansen, L. (2006). *Security as Practice: Discourse Analysis and the Bosnian War.* Abingdon: Routledge.

Hansen, N. (2013a). Derfor stemte Partii Inuit imod ophævelsen. *Sermitsiaq.* Available at: http://sermitsiaq.ag/derfor-stemte-partii-inuit-imod-ophaevelsen. [Accessed 4 September 2015].

Hansen, N. (2013b). Vi råbte hurra og sang nationalsang. *Sermitsiaq.* Available at: http://sermitsiaq.ag/kl/node/160168. [Accessed 4 September 2015].

Jacobsen, M. (2015). The power of collective identity narration: Greenland's way to a more autonomous foreign policy. *Arctic Yearbook* 4, pp. 102–118.

Jensen, C. (2013). Kuupik Kleist: Vi kan skabe miljø-problemer i 1.000 år. *Information.* Available at: https://www.information.dk/udland/2013/11/kuupik-kleist-kan-skabe-miljoe-problemer-1000-aar. [Accessed 4 September 2015].

Kielsen, K. (2013). Indlæg fra Siumut. Inatsisartut's item 106-1. Available at: http://www.inatsisartut.gl/dvd/EM2013/pdf/media/1133920/pkt106_em2013_nultolerance_ordf_1beh_siumut_dk.pdf. [Accessed 4 September 2015].

Klarskov, K. (2014a). Et radioaktivt valgtema. *Politiken* (23 November 2014), p. 7.

Klarskov, K. (2014b). Grønlandsk toppolitiker vil blæse på resultatet af en folkeafstemning. *Politiken* (23 November 2014), p. 5.

Kleist, K. (2014). Indlæg fra Inuit Ataqatigiit. Inatsisartut's item 169-1. Available at: http://www.inatsisartut.gl/dvd/FM2014/pdf/media/1808545/pkt169_fm2014_atomfritarktis_ordf_1beh_inuit_ataqatigiit_dk.pdf. [Accessed 4 September 2015].

Kruse, K. (2014a). An elemental debate. *The Arctic Journal.* Available at: http://arcticjournal.com/oil-minerals/1127/elemental-debate. [Accessed 4 September 2015].

Kruse, K. (2014b). SIK: Lad uran være til gavn. *Sermitsiaq.* Available at: http://sermitsiaq.ag/sik-lad-uran-gavn. [Accessed 4 September 2015].

Kruse, K. (2014c). Uran deler fortsat Siumut og Inuit Ataqatigiit. *Sermitsiaq.* Available at: http://sermitsiaq.ag/uran-deler-fortsat-siumut-inuit-ataqatigiit. [Accessed 4 September 2015].

Lanteigne, M. (2014). The Greenland vote: What next for international co-operation? *The Arctic Journal.* Available at: http://arcticjournal.com/opinion/1171/greenland-vote-what-next-international-co-operation. [Accessed 4 September 2015].

Lindkvist, A. (2014). Seeking stability through change. *The Arctic Journal.* Available at: http://arcticjournal.com/politics/1067/seeking-stability-through-change. [Accessed 4 September 2015].

Lund, K. (2013a). Indlæg fra Inuit Ataqatigiit. Inatsisartut's item 106-1. Available at: http://www.inatsisartut.gl/dvd/EM2013/pdf/media/1133994/pkt106_em2013_nultolerance_ordf_1beh_inuit_ataqatigiit_dk.pdf. [Accessed 4 September 2015].

Lund, K. (2013b). Indlæg fra Inuit Ataqatigiit. Inatsisartut's item 106-2. Available at: http://inatsisartut.gl/dvd/EM2013/pdf/media/1151847/pkt106_em2013_nultolerance_uran_ordf_2beh_inuit_ataqatigiit_dk.pdf. [Accessed 4 September 2015].

Mair, J. (2014). The Kvanefjeld of the future: new, attractive jobs. *The Arctic Journal.* Available at: http://arcticjournal.com/opinion/1138/kvanefjeld-future-new-attractive-jobs. [Accessed 4 September 2015].

Meinecke, S. (2014). Jakob Janussen til dansk avis: IA kan sejre ad helvedes til. *Kalaallit Nunaata Radioa.* Available at: http://knr.gl/da/nyheder/jakob-janussen-til-dansk-avis-ia-kan-sejre-ad-helvedes-til. [Accessed 4 September 2015].

Mølgaard, N. (2013a). Hans Enoksen fastholder uran-udtalelse. *Sermitsiaq.* Available at: http://sermitsiaq.ag/enoksen-fastholder-uran-udtalelse. [Accessed 4 September 2015].

Mølgaard, N. (2013b). Kirkegaard: Misforstået blanding af uranrapport og urandebat. *Sermitsiaq.* Available at: http://sermitsiaq.ag/kirkegaard-misforstaaet-blanding-uranrapport-urandebat. [Accessed 4 September 2015].

Mølgaard, N. (2013c). NGO'er: Udsæt beslutningen om nultolerancen. *Sermitsiaq.* Available at: http://sermitsiaq.ag/ngoer-udsaet-beslutningen-nultolerancen. [Accessed 4 September 2015].

Mølgaard, N. (2013d). Ny urandebat i Narsaq. Sermitsiaq. Available at: http://sermitsiaq.ag/ny-urandebat-i-narsaq. [Accessed 4 September 2015].

Nielsen, T. (2014). Uran afgjorde koalitionsforhandlingerne. *Sermitsiaq.* Available at: http://sermitsiaq.ag/uran-afgjorde-koalitionsforhandlingerne. [Accessed 4 September 2015].

Olsen, J. (2013). Indlæg fra Inuit Ataqatigiit. Speech in the Danish Parliament. Available at: http://ia.gl/da/folketingets-abningsdebat-johan-lund-olsen/. [Accessed 4 September 2015].

Petersen, G. (2013). Indlæg fra Atassut. Inatsisartut's item 88-1. Available at: http://www.inatsisartut.gl/dvd/EM2013/pdf/media/1136757/pkt88_em2013_nultolerancen_atassut_ordf_1beh_dk.pdf. [Accessed 4 September 2015].

Rosing, M. (2014). *Til gavn for Grønland*. Copenhagen: University of Copenhagen.

Sermitsiaq. (2013). Aleqa: Danmark skal blande sig udenom. Available at: http://sermitsiaq.ag/aleqa-danmark-blande-udenom. [Accessed 4 September 2015].

Steinberg, P., Tasch, J., and Gerhardt, H. (2014). *Contesting the Arctic: Rethinking Politics in the Circumpolar North*. London: I.B.Tauris.

Rottbøll, E. (2013). Grønland vil selv føre udenrigspolitik. *Sermitsiaq*. Available at: https://www.information.dk/indland/2013/10/groenland-foere-udenrigspolitik. [Accessed 4 September 2015].

Thomasen, G. (2014). Nultolerance: En politik der aldrig var? *Ren Energi* (138), pp. 1–3.

Thorning-Schmidt, H. (2013). Tale ved Folketingets åbning den 1. oktober 2013. Danish Prime Minister's Office. Available at: http://stm.dk/_p_13927.html. [Accessed 4 September 2015].

Thorsen, R. (2013). IA jubler over Thornings uranudmelding – S raser. *Sermitsiaq*. Available at: http://sermitsiaq.ag/ia-jubler-thornings-uranudmelding-s-raser. [Accessed 4 September 2015].

Thorsen, R. and Djørup, C. (2013). Ny dansk-grønlandsk arbejdsgruppe om uran. *Sermitsiaq*. Available at: http://sermitsiaq.ag/ny-dansk-groenlandsk-arbejdsgruppe-uran. [Accessed 4 September 2015].

Vestergaard, C. (2014). Greenland's uranium and the Kingdom of Denmark. In: H. Mouritzen and N. Hvidt, ed., *Danish Foreign Policy Yearbook*. Copenhagen: Danish Institute for International Studies, pp. 51–75.

Vestergaard, C. and Thomasen, G. (2015). *Governing Uranium in the Danish Realm*. Copenhagen: Danish Institute for International Studies.

Wæver, O. (2002). Identity, communities and foreign policy: Discourse analysis as foreign policy theory. In: L. Hansen and O. Wæver, eds, *European Integration and National Identity: The Challenge of the Nordic States*. Abingdon: Routledge, 20–49.

4 The Arctic turn

How did the High North become a foreign and security policy priority for Denmark?

Jon Rahbek-Clemmensen

In October 2006, representatives of the Arctic governments met in Salekhard in northern Siberia for the biennial Arctic Council ministerial meeting to discuss how the council could combat regional climate change, among other issues. While most capitals were represented by their foreign minister, a few states – Canada, Denmark, and the United States – sent other representatives. There was nothing unusual about the absence of Per Stig Møller, the Danish foreign minister – a Danish foreign minister had only once attended an Arctic Council ministerial meeting (Arctic Council 2016). Møller's nonappearance did, however, betray the low status that Arctic affairs had in the halls of government in Copenhagen. Since the end of the Cold War, where Greenland had helped tie Denmark and the US closer together due to its geostrategically important position between North America and the Soviet Union, Arctic and Greenlandic affairs had mainly been about managing fishing quotas. Though crucial for Danish–Greenlandic relations, such issues were hardly top priorities for Her Majesty's Foreign Service.

In April 2014, less than eight years after the Salekhard ministerial, Martin Lidegaard, then the Danish minister of foreign affairs, highlighted the Arctic as one of his four foreign policy priorities in a speech to the Danish Foreign Policy Society. The High North, the minister argued, was "increasingly an arena for global political and economic forces. ... Denmark has played and must play a special role there" (Lidegaard 2014a). The speech was just one in a long line of new polar initiatives – which included an Arctic strategy, a new Self-Government Act that gives Greenland a roadmap to independence, a restructuring of the defence institutions in the North Atlantic, new investments in modern patrol vessels, and discussions about possible satellite surveillance cooperation with Canada – presented by Lidegaard and his predecessors.

The difference between 2006 and 2014 illustrates that Denmark's foreign, defence, and security policy underwent an *Arctic turn,* as Danish politicians, experts, media, and civil servants have directed their attention towards Greenland and the High North. By the Arctic turn I mean a shift in policy focus, where policymakers come to view Denmark and Greenland as part of a wider region driven by unique economic, political, and social forces that warrant special attention from government agencies. Scholars at the time were quick to notice that a new foreign and security policy arena was opening and several studies have mapped the challenges facing Denmark in the Arctic (Petersen 2008; 2009; 2011; Rahbek-Clemmensen et al. 2012; Rasmussen 2013; Rahbek-Clemmensen 2014). These studies have focused on the *external challenges* and one can easily be led to believe that the Arctic turn happened almost automatically – that the relevant ministries recognised a new challenge and quickly devised strategies and policies to handle that challenge, based on an already existing understanding of the strategic situation in the region. However, the fact that the minister for foreign affairs stayed away from an Arctic Council ministerial less than two years before Denmark would play a pivotal regional role by hosting the 2008 Ilulissat meeting shows that the Arctic turn was a learning process. The literature on organisational change highlights that organisations are slow to adapt to new policy challenges and that different institutions favour different policy options based on their own function in the national policy machinery (Allison 1969, 698–707).

The present chapter examines how the Danish Ministry of Foreign Affairs (MFA) and Ministry of Defence (MoD) changed how they thought about the High North between 2005 and 2015 and how the Arctic turn leads to policy challenges for Danish and Greenlandic policymakers. It took some time for the Danish ministries to understand the new challenges they were facing in the Arctic and the different institutions did not adapt similarly, although the discrepancies between them were rather small. The Arctic turn began around the middle of the 2000s within the MFA, but it was originally mainly a *Greenlandic* turn – that is, it focused on improving relations with Greenland, but paid little attention to the wider Arctic. The Arctic truly came on the agenda around 2008–11 and both the MFA and the MoD contributed to this change. The Arctic turn entailed a shift in strategic priorities and *modus operandi* as well as a shift in geographical focus – Denmark recognised that it had an interest in the status quo and developed an Arctic strategy that focused on furthering regional cooperation through accommodation and diplomacy. The Danish Armed Forces (DAF) became a tool in that strategy, as it could improve Denmark's relationship to other states and the government of Greenland through practical initiatives and cooperation. Danish policymakers struggle to find the balance between cooperation and conflict/ competition (i.e. should Denmark further cooperation or prepare for eventual conflict or competition?) and between diplomacy and operational effectiveness (i.e. should the DAF cooperate with others to improve effectiveness only or do the diplomatic effects of practical cooperation make it a good in of itself?). However, perhaps most importantly, the Arctic turn decreases Greenland's room for manoeuvre regionally and within the Kingdom of Denmark, as Copenhagen has become more interested in regional diplomacy and therefore pays less attention to the bilateral relationship with Nuuk.

The analysis tracks when and how the Arctic turn unfolded, based on in-depth analysis of written sources (public foreign policy and defence reports, agreements, strategies, and statements) and interviews with current and former civil servants and experts. Organisational learning can be seen in instances where the two ministries were slow to adapt to the new changes and where they espoused different strategic understandings and priorities. The chapter focuses specifically on foreign and security policy and it consequently focuses on the two ministries – the MFA and MoD – where policy changes are most likely to occur. It leaves aside other relevant ministries, such as the Prime Minister's Office and the Ministry of Finance, as these institutions primarily focus on day-to-day coordination rather than long-term strategic thinking. The chapter focuses specifically on the period between 2005 and 2015, although it draws lines back to previous decades.

The argument of the chapter will progress in three steps. The first section examines shows how the MFA became interested in the Arctic and how its early thinking focused on the bilateral relationship between Denmark and Greenland. Having done that, the following section shows how the focus of Danish policymakers shifted from Greenland to the wider Arctic region and how they came to develop a cooperation-oriented regional strategy. The final section examines how this shift affected defence planning and the obstacles involved in strategic coordination between the MFA and the MoD.

A renewed focus on Greenland (2005–2008)

The Arctic turn began in 2007 as a reaction to Russia's demonstrative planting of a titanium flag on the North Pole seabed. Of course, Danish ministries, including the MFA and MoD, had always dealt with Greenlandic and Faroese matters, but they had been areas of rather marginal importance. For instance, a 2003 foreign policy white book, which outlined Danish foreign policy priorities in the wake of 9/11, did not mention the Arctic at all and only mentioned Greenland in passing (Government of Denmark 2003). For individual diplomats, an appointment to the Greenland and Faroe Islands desk did not build the foundation for a promising career (interview with a former civil servant from the Ministry of Foreign Affairs (A), February 2016; interview with a former civil servant from the Ministry of Foreign Affairs (B), February 2016; interview with a former civil servant from the Ministry of Foreign Affairs, March 2016). Greenland had, of course, played an important role in Danish foreign policy during the Second World War and the Cold War, where the island's strategic importance for the United States had made it an important factor in Copenhagen's transatlantic relations (Danish Foreign Policy Institute 1997). Since Greenland got home rule in 1979, Denmark had had to manage Greenland's at times complicated foreign relations without disregarding or affronting Nuuk. In that sense, Arctic matters were not new to the MFA, but the focus was primarily on Danish–Greenlandic relations and not on the Arctic as a region with its own political dynamics (interview with a former civil servant from the Ministry of Foreign Affairs (A), February 2016; interview with a former civil servant from the Ministry of Foreign Affairs (B), February 2016; interview with a former civil servant from the Ministry of Foreign Affairs, March 2016).

Similarly, strategic thinking about the Arctic did not begin from scratch in the MoD. One can recognise certain nascent geo-economic and defence diplomacy thinking going back to the 1990s, but the Arctic and Greenland generally became a main priority from around 2008–2009. Danish defence and security thinking before the late 2000s paid little attention to Greenland and mainly understood the island as an operational area within Denmark, where the DAF (mainly the navy) had to be present and provide services without much consideration for national or international diplomacy. As Figures 4.1 and 4.2 show, Greenland and the Arctic were almost completely absent from Defence Commission reports and the five-year defence agreements, the two documents in which the main priorities of the broad consensus within parliament are described and concrete decisions are made, and they mainly appeared in relations to minor technical discussions about Island Command Greenland and Island Command Faroes (Danish Parliament 1999; 2004; 2009; 2012; Defence Commission 2009).[1]

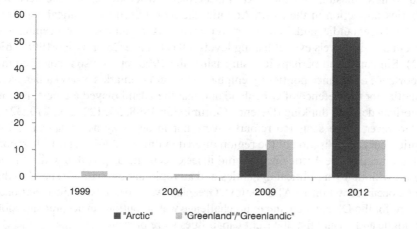

Figure 4.1 Use of terms in Danish Defence Agreements, including appendices (1999–2012)

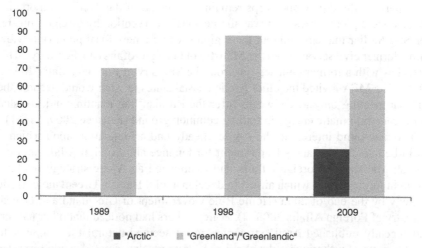

Figure 4.2 Use of terms in Danish Defence Commission Reports, excluding appendices (1989–2009)

Greenland was not wholly missing from defence and security thinking in that period. The report of the 1997 Defence Commission (which was published in 1998) discussed several questions related to Greenland and even included certain embryonic elements of what would later become the Arctic turn.[2] When considering factors that could expand the mission load for the navy, it expected "increased activities in relation to the search for and exploitation of resources in the Greenlandic and Faroese shelf area in coming years, which can perhaps actualise questions of territorial delimitation" (Defence Commission 1998, 113). In comparison, such thoughts had been completely absent from the 1988 Defence Commission report (Defence Commission 1989). Policymakers were aware that geo-economic changes could have geopolitical impacts for Denmark and lead to new missions for the DAF. Furthermore, when discussing the need for fisheries inspection in the North Atlantic, the report used the political concerns of the Greenlandic and Faroese governments as arguments for keeping the current activity levels even if fishing levels fell (Defence Commission 1998, 263, 282). Similarly, and perhaps less surprising, the 1998 report also recognised the importance of alliance politics by emphasising that Denmark's NATO allies were essential for the defence of Greenland and that the island played a crucial role in American defence thinking (Defence Commission 1998, 30, 122, 259, 270, 321).

However, these scattered remarks were not in any way as advanced as the explicit conceptualisation of the region offered in the late 2000s. The 1998 report did not mention the Arctic once nor did it see Greenland as part of a wider Arctic region with its own political dynamics and all diplomatic relations related to the island occurred within a NATO context. Greenland was an almost static operational theatre for the DAF, where the main challenge was to outline an accurate mission catalogue and ensure that the right capabilities were present.

The 2007 Russian flag planting caught the attention of the MFA, but only due to direct intervention by the political level. According to several former diplomats, large parts of the diplomatic corps remained sceptical of the Arctic's importance and, as one diplomat recalls, it was the personal intervention by foreign minister Per Stig Møller that ensured that the region became a new focal point (interview with a former civil servant from the Ministry of Foreign Affairs (A), February 2016; interview with a former civil servant from the Ministry of Foreign Affairs, March 2016). The MFA invited the other Arctic coastal states to what would become the Ilulissat meeting only a few weeks after the Russian flag planting and invested significant diplomatic energy in finding common ground (Petersen 2009, 54–61).

The newfound interest in the Arctic already had a foundation upon which it could be built. Greenland's Department for Finance and Foreign Affairs and the Danish Ministry of Foreign Affairs had established an Arctic strategy working group in August 2006, which aimed to develop a joint Danish–Greenlandic Arctic strategy by the end of 2006 (Home Rule Government of Greenland and Danish Ministry of Foreign Affairs 2008, 4). Policymakers had noticed that other nations had recently published Arctic strategies and it seemed natural for Denmark to do the same. Furthermore, developing a strategy also created a platform for improving relations with Nuuk as Greenland moved towards a looser constitutional relationship with Copenhagen with the new independence agreement (which was finalised in 2009) (interview with a former civil servant from the Ministry of Foreign Affairs (A), February 2016). This initial timeline soon turned out to be too optimistic and the working group only published a strategy draft – *The Arctic in Times of Change* – in May 2008, as a prelude to the Ilulissat ministerial.

Despite being an attempt to formulate an *Arctic* strategy and despite being published simultaneously with the Ilulissat Declaration (where Denmark made its diplomatic mark on the region), the 2008 draft was surprisingly inward-looking and barely addressed Arctic politics. It represented the prevailing focus on Greenland among the MFA's Arctic officials. It dedicated a lot of space to the possibility and desirability of Greenlandic independence and it generally reads as a Danish attempt at improving relations with Greenland rather than an Arctic strategy. It only scantly touched upon how and why the Arctic was changing and only mentioned that climate change and globalisation caused "big changes in both activity levels and living conditions" (Home Rule Government of Greenland and Danish Ministry of Foreign Affairs 2008, 4) and affected businesses such as hydropower, shipping, and hydrocarbon industries. It almost entirely focused on the repercussions for *Greenland* and how new initiatives and cooperation between Copenhagen and Nuuk could open new opportunities there (Home Rule Government of Greenland and Danish Ministry of Foreign Affairs 2008, 21–24, 33–34). Similarly, the draft did not address the political dynamics of the Arctic (though it did provide an overview of existing regional institutions and initiatives) nor did it present significant new initiatives to improve Danish interests or relations to other states in the region at large (Home Rule Government of Greenland and Danish Ministry of Foreign Affairs 2008, 13–18). It only provided an overview of current DAF activities and left any discussions about the future of the DAF to the upcoming 2008 Defence Commission, although it did suggest that the DAF should increase the inclusion of Greenlanders in its activities in Greenland (Home Rule Government of Greenland and Danish Ministry of Foreign Affairs 2008, 11–13). This idea, which resembled schemes which had been tried out since the 1950s, was clearly another tool to improve relations between Nuuk and Copenhagen (Jensen 2001, 142).

In sum, the period between 2005 and 2008 illustrates that Danish adaptation to the new challenges in the Arctic did not occur automatically. The 2007 Russian flag planting meant that the MFA came to pay more attention to the High North, but it did not fully realise the scope of the new challenges and it stayed within a Greenland-centric understanding of the region in at least the following year. It is remarkable the MFA line of thinking did not change even as it organised the Ilulissat meeting, which came to play a pivotal role for multilateral governance in the region.

From Greenland to the Arctic (2008–2015)

The period after the Ilulissat meeting was marked by a rapid shift in strategic thinking about the Arctic. Key publications revealed an increased awareness of the intricacies of Arctic politics and the emphasis shifted from Greenland to the Arctic more generally speaking. The MoD was included in strategic thinking as lofty strategic goals were made concrete through specific programmes and initiatives.

The 2008 Defence Commission's report (published 2009) was the first major publication to show this shift. It was bound by its mission statement to analyse "the security situation as it is being influenced by the development in the Arctic areas" and it shows that the ministries' focus was slowly moving from Greenland to the wider Arctic (Defence Commission 2009, 17). The commission mainly focused on other priorities, such as finding new funds for the Afghanistan mission and the pending fighter aircraft procurement process, and very little time was spend on discussing the Arctic (interview with a former member of the 2008 Defence Commission, February 2016).

The sections about Arctic politics were most likely produced within the MFA and they presented a more sophisticated view of the Arctic than the one found in the strategy draft from the previous year. Where the MFA strategy draft noted that a change was occurring, but only vaguely pointed to climate change and globalisation as causes, the defence commission report identified specific factors (global energy trends and melting sea ice) as the main drivers of regional change (Defence Commission 2009, 71). It also went further in characterising the political climate in the region, which it saw as defined by both cooperation and competition. On the one hand, it mentioned that the states had agreed to delimit the region according to international law in the Ilulissat Declaration (which had been made shortly after the publication of the 2008 strategy draft). On the other hand, it foresaw increased competition as the regional states, especially Canada, Norway, and Russia, had "started to position themselves to maximise their starting point in relation to the expected development" (Defence Commission 2009, 71–72). This regional competition had been absent from the 2008 strategy draft and would also be missing from the 2011 Arctic strategy.

The 2011 *Strategy for the Arctic 2011–2020* represented the epitome of the Arctic turn as the narrow focus on Danish–Greenlandic relations was replaced by a vision of a united Kingdom of Denmark (Denmark proper, the Faroe Islands, and Greenland) in a rapidly changing region. Making the kingdom the acting subject of the strategy entailed different political priorities. Gone were the discussions of Greenlandic independence as the strategy assumed and worked for the longevity of the current constitutional arrangement.

The strategy offered a detailed understanding of Arctic politics, which was seen as defined by interstate cooperation while recognising that future conflict was possible. The Arctic order depends on international law and continued and enhanced cooperation on practical issues – principles that were solidified in the 2008 Ilulissat Declaration. The kingdom had an interest in the status quo and should play "a key role in shaping the future international architecture of the Arctic" by continuing this work to preserve regional stability (Government of Denmark et al. 2011, 50). The strategy addressed several groups of actors with the Arctic coastal states (Canada, Denmark, Norway, Russia, and the United States) and, to a lesser extent, the other Arctic Council member states (Finland, Iceland, and Sweden) as the axis of regional decision-making. The kingdom should aim to cooperate with non-Arctic states or state-like entities (e.g. the EU, China, Japan, and South Korea), and non-state actors (the strategy mentions indigenous peoples' organisations, but also seems to target environmental NGOs). Though it highlighted the need for minor accommodations to these actors' interests, it did not envision fundamental changes to the regional order (Government of Denmark et al. 2011, 32, 35, 41, 49–55).

The strategy tried to conceal a fundamental tension between regional cooperation and conflict by focusing on the former and downplay the latter. Curiously, though the strategy hinted at the possibility of regional conflict and competition, it never described how such disturbances of the regional order could come about nor did it identify potential revisionist actors. The most detailed description of the potential for conflict and competition consisted of a few lines emphasising that "[e]ven though the working relationship of the Arctic Ocean's coastal states is close, there will be a continuing need to enforce the Kingdom's sovereignty ... by the armed forces through a visible presence in the region" (Government of Denmark et al. 2011, 20). The section went on to stress that

> [w]ithin the entire spectrum of tasks, the Kingdom attaches great importance to confidence building and cooperation with Arctic partner countries ... to maintain the Arctic as a region characterised by cooperation and good neighbourliness.
>
> (Government of Denmark et al. 2011, 18, 20–21)

It also highlighted surveillance and search-and-rescue as potential collaboration areas.

The DAF should thus not just perform traditional military and coastguard missions, but should also play an active diplomatic role to create new links to other states. This new political role should not just focus on the kingdom's external partners. Following the suggestion made in the 2008 strategy draft, the DAF

> aspires ... to reflect the surrounding community. Indeed it is a Danish-Greenlandic hope that citizens of Greenland can be increasingly involved in the tasks of the armed forces and with that, participate in a wide range of training opportunities, whether they be basic training, civil/military specialist and management training programs or customised further education at all levels.
>
> (Government of Denmark et al. 2011, 21)

Though an enhanced involvement of the Greenlandic population will most likely have an operational effect (Kristensen et al. 2013; Østhagen 2017), these would be minor and it is not difficult to view the initiative as having a *political* purpose: the DAF was thus meant to serve as a political integration mechanism between Denmark and Greenland by recruiting and collaborating with the Greenlandic population.

Similarly, certain policy areas were redefined in a way that strengthened the interests of Denmark proper. For instance, compared to the 2008 strategy draft, the *Arctic Strategy* was vaguer in defining strategic priorities for the relationship between the kingdom and the EU. The 2008 strategy draft had highlighted specific programs (including the EU's Northern Dimension and the Overseas Countries and Territories programme) where Greenland could benefit from active involvement by shaping activities and priorities and attracting funding. These considerations were replaced with more generic goals, such as shaping "EU policies relevant to the Arctic and Arctic challenges, and in this context ... ensure the Arctic peoples' rights and interests" (Home Rule Government of Greenland and Danish Ministry of Foreign Affairs 2008, 15–17; Government of Denmark et al. 2011, 52–53).

Thus, the Arctic turn not just involved a geographical widened perspective from Greenland to a broader region, defined by its own political dynamic and connected to non-Arctic actors. It also entailed a new *political* approach to the region that aimed to protect the status quo through diplomacy, most importantly by accommodating other states' interests (to a certain extent) and enhancing existing cooperation. In the 2011 strategy, defence and coast guard missions were crucial in the Arctic and there was a large potential for cooperation, which meant that the DAF could become a vehicle for further collaboration. As the Arctic became defined by a changing and complex political landscape, the kingdom had to operate as a unified actor and Greenlandic independence was not addressed in the strategy. Greenland had previously used Copenhagen's disinterest in Arctic matters to acquire significant influence over the kingdom's regional policy, but this influence was transferred back to Denmark with the new international focus on the Arctic and the Arctic Strategy (interview with former civil servant from the Ministry of Foreign Affairs, March 2016). Of course, Greenland was not entirely forgotten within the halls of government in Copenhagen. Greenlandic independence remains a real option and the MFA still worked to improve Danish–Greenlandic relations by e.g. improving living conditions, attracting investments from abroad, and furthering integration in the DAF.

The Arctic turn contained tensions between cooperation and conflict and between diplomacy and operational effectiveness. The Kingdom of Denmark would work towards diminishing any tensions through accommodation, cooperation, and institutionalisation, but it concurrently had to be able to react to infringements on Danish sovereignty or a future possibility of a regional conflict. Preparing for future conflict would send a signal to other states, which could, in turn, lead to security dilemmas and increase the likelihood of conflict. The MFA thus faced a choice between emphasising conflict or cooperation and chose to focus on the latter. Similarly, it also suggested that the DAF could contribute to de-escalation by increasing defence diplomacy and military cooperation with other states and by incorporating Greenlandic communities in its operations.

During this period, it became obvious that the Arctic turn was more than mere words, as parliament decided to procure a third Knud Rasmussen class patrol vessel to replace the last, aging Agdlek cutter (an investment that had been on the table, but not decided upon, since the 1990s) (Defence Commission 1998, 284; Danish Parliament 2012, 10). It speaks volumes that both the 2010–14 and 2013–17 defence agreements – which both focused on rationalising and downsizing the DAF in the wake of the financial crisis – allocated funds for Arctic-related projects, while generally slashing funding for the DAF.

In sum, from 2008, Danish policymakers' interest in the High North continued to increase and they began to think about the Arctic as a region in flux, characterised by crucial multilateral relations between 2008 and 2011. Danish foreign and security policy thinking experienced a truly Arctic turn. The relationship to Greenland still remained crucial, but Copenhagen became aware of the importance of regional diplomacy. The 2011 Arctic strategy emphasised regional cooperation over potential conflict and it suggested concrete initiatives that aimed to engage Greenland and Arctic and non-Arctic states.

The Arctic turn within the DAF

The way Danish defence institutions adapted to the Arctic turn further illustrates that organisational learning shape thinking about High North matters. The defence institutions largely adapted to the Arctic turn, though the shift was often slow and at times incomplete. Various strategic and operational initiatives have made defence institutions much more aware of the Arctic as a focal point for Danish defence policy (interview with a civil servant from the Ministry of Defence, January 2016; interview with former civil servant from the Ministry of Foreign Affairs, March 2016). Both the 2010–14 and the 2013–17 agreements identified Denmark as a central player in a changing Arctic and highlighting the need for further cooperation with other states in the Arctic (Danish Parliament 2009, 10–11; 2012, 14, 41–42). The emphasis on the High North increased even more in the 2013–17 defence agreement, which highlighting the region as one of three crucial future challenges (the others being the drawdown of combat forces from Afghanistan and the cyber threat) (Danish Parliament 2012, 4, 7).

The defence institutions slowly adopted the MFA's line of thinking, which emphasised awareness of signals sent to other nations and that the DAF should play an active political role vis-à-vis other states and the government of Greenland. When asked directly, operational planners estimate that defence institutions became more aware of the importance of the signals they send to other Arctic nations and to the Greenlandic public (interview with civil servants from the Ministry of Defence, February 2016).

A former MFA diplomat recalls having had a good working relationship with the MoD and that defence institutions over time became more aware of the importance of political signalling. The diplomat also recalls that the MFA and MoD did emphasise different aspects of regional politics, with the latter being more concerned with traditional defence issues (interview with a former civil servant from the Ministry of Foreign Affairs, March 2016). However, it did take some time for the defence institutions to make the Arctic a new focus area and to adapt to the MFA's cooperation-oriented course. One example is the aforementioned efforts to include the Greenlandic society in defence activities. The 2008 Defence Commission report and the following 2010–14 defence agreement from 2009 (in which actual funds for project were allocated) omitted the possibility of further involvement of Greenland society in its activities, even though this idea had been floated in the 2008 strategy draft. The idea was only adopted by the MoD in 2012, after it had been explicitly mentioned in the 2011 Arctic strategy (Danish Parliament 2012, 14). The MoD thus hesitated to adopt a mission, which served a political rather than an operational purpose.

The MoD was slow to become aware of the signals it sent to other nations, but it eventually adopted a cooperation-oriented approach to the region akin to the one coming out of the MFA (interview with a civil servant from the Ministry of Defence, January 2016; interview with civil servants from the Ministry of Defence, February 2016; interview with a former civil servant from the Ministry of Foreign Affairs, March 2016). For instance, several interviewees highlight the establishment and subsequent renaming of the Arctic Task Force (Arktisk Indsatsstyrke) as an example of the MoD's lateness. The Arctic turn forced the DAF to scale up its search-and-rescue and environmental protection capacities, but budget constraints and the aforementioned focus on Afghanistan compelled defence planners to do so on the cheap. The solution was to create a programme in 2009 – the Arctic Task Force – where existing units were trained and prepared to be inserted into an Arctic theatre in case of an emergency (Danish Parliament 2009, 10–11; Ejsing 2009; interview with a civil servant from the Ministry of Defence, January 2016). However, though operationally sensible, the program had unfortunate and unforeseen political connotations. Its English name made it sound like a military force and observers from other states interpreted it as an example of Danish assertiveness in the High North (BBC 2009; Huebert 2010, 10–12; Rahbek-Clemmensen et al. 2012, 28). The MoD changed it to the more pacific Arctic Preparedness Force (Arktisk Beredskabsstyrke) when it became aware of the name's counterproductive associations, which thus shows that it was willing to take signalling seriously, but that it took some time to adapt to the new realities (interview with a civil servant from the Ministry of Defence, January 2016; interview with civil servants from the Ministry of Defence, February 2016; interview with a former civil servant from the Ministry of Foreign Affairs, March 2016).

The Danish Defence Intelligence Service's (DDIS) annual risk assessments provide another example of discrepancies between defence institutions and the MFA. It is remarkable that neither the Russian flag planting on the North Pole, the Ilulissat Declaration, nor the MFA's 2008 strategy draft led the DDIS to address the region. Only in 2009, after the publication of the 2008 defence commission report, did the *Intelligence Risk Assessment* include a half-page sub-section on "the development in the Arctic area" and this analysis betrayed a rather crude understanding of the region (Danish Defence Intelligence Service 2009; 20). For instance, the report argued that

> because of the natural resources that may be exploitable in the Arctic Ocean in the future, it is likely that opposing interests between the [Arctic] coastal state will appear. This will increase the demands to the UN's Continental Shelf Commission [sic] as a forum for possible conflict solution.
> (Danish Defence Intelligence Service 2009, 20)

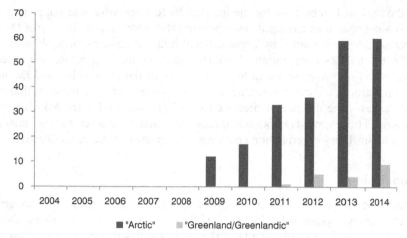

Figure 4.3 Use of terms in DDIS Intelligence Risk Assessments (2004–2014)

Of course, the UN Commission for Delimitation of the Continental Shelf (CLCS), to which the report referred, only evaluates the scientific validity of continental shelf claims and it is not a forum for conflict resolution. The DDIS was thus slow to make the High North a new focus area and its analysts were unfamiliar with the political dynamics of the region.

Generally speaking, the DDIS's *Intelligence Risk Assessments* offered a bleaker picture of Arctic politics than the one presented by the MFA and it highlighted the potential for competition between the states in the region (Danish Defence Intelligence Service 2009, 20; 2010, 40). The DDIS warned that Chinese investments in just a few industrial projects could make up a sizable portion of the Greenlandic economy and thus lead to greater Chinese influence in Greenland (Danish Defence Intelligence Service 2012, 11–12; 2013, 13–15; 2014, 29–31). It also pointed out that global tensions between Russia and the US could spill over into the Arctic, even though the geopolitical make-up of the Arctic supported cooperation, even before that actually partly occurred in the wake of the 2014 Ukraine crisis (Danish Defence Intelligence Service 2014, 15–16). These dynamics were absent from the 2011 Arctic strategy and the Minister of Foreign Affairs' annual *Statement about Arctic Cooperation* before the Ukraine crisis (Government of Denmark et al. 2011; Søvndal 2013). After the crisis had begun, where the DDIS warned that "increasing Russian determination to take unilateral actions to reach its strategic objectives at the expense of international relations could also extend to the Arctic", Martin Lidegaard, the then-Minister of Foreign Affairs told parliament that "there are no indications that Russia wants to bring the tensions of the Ukraine crisis to the Arctic cooperation" and the possibility of conflict spill-over from Ukraine into the Arctic was absent from his annual Security Policy Assessment (Danish Defence Intelligence Service 2014, 29; Lidegaard 2014b).[3]

In sum, the defence institutions' reactions to the Arctic turn further illustrate that Danish institutions did not adapt automatically to the new developments in the region. It took time for the institutions to learn what was happening and seeing the region as having its own political dynamic. It also shows that Danish Arctic thinking is shaped by organisational functions and cultures. Whereas the MFA emphasises cooperation, diplomacy, and political signalling, the defence institutions are more prone to highlight the potential for conflict and the need for operational effectiveness. One should not overemphasise these institutional differences – the defence institutions have largely adapted to the MFA's way of thinking. However, it is important to remember that these discrepancies exist and that civil-military coordination could become an issue in the years to come.

Conclusion

The Arctic turn entailed a learning-process for the Danish ministries. Though the Arctic turn was precipitated by the 2007 Russian flag planting, it took the Danish ministries a while to understand that the High North was becoming a region with its own political dynamic. The MFA was the first institution to shift its policy focus north, but the organisation hesitated to adopt what they had so far thought to be a rather marginal interest area. The first polities focused on Danish-Greenlandic relations and it was generally more of a *Greenlandic* turn. The MFA and MoD both contributed to the development of a cooperation-oriented strategy (codified in the 2011 Arctic Strategy), which stressed the potential for engagement and de-escalation. The DAF largely adapted to this strategy and became more aware of its signalling vis-à-vis other nations and its political role in Greenlandic society. However, adaptation was not unproblematic. It took time and effort to spread awareness of the cooperation-oriented strategy and there are minor tensions between certain institutions.

Denmark's regional diplomacy will continue to work to find a balance between regional cooperation and conflict. In the 2011 strategy, the MFA tried to paper over the tension by emphasising cooperation throughout the document, but other agencies have been willing to highlight the potential for conflict. This tension is thus also a civil-military one. While the MFA is eager to emphasise the importance of cooperation with other Arctic nations, at least some parts of the MoD see a larger potential for conflict in the region. The analysis showed a widespread consensus between the MFA and most of the MoD, but it is possible that this consensus may not continue in the future, which would cause friction between the different agencies. Similarly, the ministries must also overcome a tension between political considerations and operational effectiveness. This tension also overlaps with the civil–military divide, as the DAF tends to be more focused than the MFA on operational effectiveness. These differences may lead to different estimates of the utility of specific programmes that concurrently serve a diplomatic and an operational purpose, such as the DAF's military cooperation with the armed forces and coastguards of other Arctic states and the recruitment programmes that aim to engage Greenlanders in the DAF.

However, perhaps the most important consequence of the Arctic turn is its implications for Danish–Greenlandic relations within the Kingdom of Denmark.

The Arctic turn entailed a shift in focus away from Danish–Greenlandic relations towards building bi- and multilateral cooperation with other states in the region. This move creates both an opportunity and a challenge for Greenland. Danish policymakers have begun to look north as the Arctic has become a foreign policy priority for Denmark and this can be converted to investments and policy initiatives in Greenland. Attention also means that Danish foreign policy strategising becomes clearer. The 2011 Arctic Strategy sharpened Denmark's approach to a host of questions and it enables Danish diplomats to take advantage of new opportunities – say foreign investment opportunities – that may benefit Greenland.

The Arctic turn concurrently shifts Denmark's foreign policy priorities away from purely Greenlandic interests. Danish and Greenlandic interests do not always overlap and the Arctic turn creates a framework that allows Denmark to prioritise its interests in the name of the kingdom. For example, in the 2011 Arctic Strategy, the relationship to the EU serves as a case in point. Like Greenland, Denmark is, of course, interested in attracting specific funding from EU programmes and shaping the EU's High North policy, but Copenhagen also sees its unique Arctic position as a wedge that can open doors in Europe (in which Nuuk is less interested). Danish diplomats have to balance several concerns and need strategic flexibility to take advantage of developments in Europe. They cannot be tied down by too concrete promises. Consequently, the 2011 Arctic Strategy's sections on the EU were thus less specific and more generic than the ones found in the 2008 strategy draft. Greenland's interests are thus crowded out by the enhanced Danish awareness. In that sense, the opening of the Arctic has both positive and negative consequences for Greenland: the world's attention may be moving northward, but with it comes an enhanced Danish interest that narrows Greenland's room for manoeuvre, both regionally and within the Kingdom of Denmark.

Acknowledgements

The author would like to thank the many experts and civil servants (former and current) who contributed with their account of the events covered in this chapter. He would also like to thank Danita Burke, Chiara de Franco, Ulrik Pram Gad, Marc Jacobsen, Peter Viggo Jakobsen, Uffe Jakobsen, Vincent Keating, Kristian Søby Kristensen, Mikkel Runge Olesen, Sten Rynning, Olivier Schmitt, and Camilla Sørensen for their helpful feedback and comments. The research for this chapter was funded by the Carlsberg Foundation.

Notes

1 The 2013 agreement was made in the middle of an existing agreement. When quantifying the use of terms, one has to take into account that defence agreements have become longer and hence the terms are more likely to appear in later agreements. However, for the sake of this argument, the trend seems so strong that one can disregard such caveats.

2 One should not exaggerate the fact that variations of the term "Greenland" are mentioned more often in the earlier reports. The 1997 report was significantly longer than the 2009 report and thus more likely to mention the terms.
3 The MFA changed its course slightly the following year, when it recognised that "the Ukraine crisis has so far had a limited impact on Arctic cooperation" and that "the development of international relations outside of the region affects the security situation in the Arctic" (Jensen 2015).

References

Allison, G. (1969). Conceptual models and the Cuban missile crisis. *The American Political Science Review* 63(3), pp. 689–718.

Arctic Council. (2016). All Arctic Council Declarations 1996–2015. Available at: https:// oaarchive.arctic-council.org/bitstream/handle/11374/94/EDOCS-1200-v3-All_Arctic_ Council_Declarations_1996-2015_Searchable.PDF. [Accessed 7 Nov 2016].

BBC. (2009). Denmark plans forces for Arctic. Available at: http://news.bbc.co.uk/2/hi/ europe/8154181.stm. [Accessed 31 January 2016].

Danish Defence Intelligence Service. (2009). *Efterretningsmæssig Risikovurdering 2009*. Copenhagen: Danish Defence Intelligence Service.

Danish Defence Intelligence Service. (2010). *Efterretningsmæssig Risikovurdering 2010*. Copenhagen: Danish Defence Intelligence Service.

Danish Defence Intelligence Service. (2012). *Efterretningsmæssig Risikovurdering 2012*. Copenhagen: Danish Defence Intelligence Service.

Danish Defence Intelligence Service. (2013). *Efterretningsmæssig Risikovurdering 2013*. Copenhagen: Danish Defence Intelligence Service.

Danish Defence Intelligence Service. (2014). *Efterretningsmæssig Risikovurdering 2014*. Copenhagen: Danish Defence Intelligence Service.

Danish Foreign Policy Institute. (1997). *Grønland under den Kolde Krig, Dansk og Amerikansk Sikkerhedspolitik 1945-68*. Copenhagen: Danish Foreign Institute.

Danish Parliament. (1999). *Forsvarsforlig 2000–2004*. Copenhagen: Ministry of Defence.

Danish Parliament. (2004). *Forsvarsforlig 2005–2009*. Copenhagen: Ministry of Defence.

Danish Parliament. (2009). *Forsvarsforlig 2010–2014*. Copenhagen: Ministry of Defence.

Danish Parliament. (2012). *Forsvarsforlig 2013–2017*. Copenhagen: Ministry of Defence.

Defence Commission. (1989). *Forsvaret i 90'erne. Beretning fra Forsvarskommissionen af 1988*. Copenhagen: Ministry of Defence.

Defence Commission. (1998). *Fremtidens Forsvar. Beretning fra Forsvarskommissionen af 1997*. Copenhagen: Ministry of Defence.

Defence Commission. (2009). *Dansk forsvar: Globalt engagement. Beretning fra Forsvarskommissionen af 2008*. Copenhagen: Ministry of Defence.

Ejsing, J. (2009). Danmark opruster i Arktis. *Berlingske Tidende* (15 July 2009), p. 12.

Government of Denmark. (2003). *En verden i forandring – Regeringens bud på nye prioriteter i Danmarks udenrigspolitik*. Copenhagen: Ministry of Foreign Affairs.

Government of Denmark, Government of Greenland, and Government of the Faroe Islands. (2011). *Strategy for the Arctic 2011–2020*. Copenhagen: Ministry of Foreign Affairs.

Home Rule Government of Greenland and Danish Ministry of Foreign Affairs. (2008). *Arktis i en Brydningstid. Forslag til Strategi for Aktiviteter i det Arktiske Område*. Copenhagen: Ministry of Foreign Affairs.

Huebert, R. (2010). *The Newly Emerging Arctic Security Environment*. Calgary: The Canadian Defence & Foreign Affairs Institute.

Jensen, K. (2015). Redegørelse af 8/10 15 om samarbejdet i Arktis. Speech given at the Danish Parliament. Available at: http://www.ft.dk/samling/20151/redegoerelse/r3/1553595.pdf. [Accessed 4 November 2015].

Jensen, P. (2001). *Grønlands Kommando i 50 år*. Copenhagen: Aschehoug.

Kristensen, K., Hoffmann, R., and Pedersen, J. (2013). *Samfundshåndhævelse i Grønland: Forandring, Forsvar og Frivillighed*. Copenhagen: Center for Military Studies.

Lidegaard, M. (2014a). Verden set fra Danmark. Speech given at the Danish Foreign Policy Society on 10 April 2014. Available at: http://um.dk/da/om-os/ministrene/tidligere-ministres-taler-og-artikler/martin-lidegaard-taler-og-artikler/verden-set-fra-danmark/. [Accessed 16 February 2016].

Lidegaard, M. (2014b). Redegørelse af 9/10 14 om arktisk samarbejde. Speech given at the Danish Parliament on 9 October 2014. Available at: http://www.ft.dk/samling/20141/redegoerelse/r3/1407178.pdf. [4 November 2015].

Østhagen, A. (2017). *Utilising Local Capacities: Maritime Emergency Response across the Arctic*. Copenhagen: Center for Military Studies.

Petersen, N. (2008). Truslen i nord. In: H. Mortensen, ed., *Helt Forsvarligt? Danmarks Militære Udfordringer i en Usikker Fremtid*. Copenhagen: Gyldendal, pp. 91–108.

Petersen, N. (2009). The Arctic as a new arena for Danish foreign policy: The Ilulissat initiative and its implications. In: H. Mouritzen and N. Hvidt, ed., *Danish Foreign Policy Yearbook 2009*. Copenhagen: Danish Institute for International Studies, pp. 35–78.

Petersen, N. (2011). The Arctic challenge to Danish foreign and security policy. In: J. Kraska, ed., *Arctic Security in an Age of Climate Change*. Cambridge: Cambridge University Press, pp. 145–165.

Rahbek-Clemmensen, J. (2014). 'Arctic-vism' in practice: The challenges facing Denmark's political-military strategy in the High North. *Arctic Yearbook* 3, pp. 399–414.

Rahbek-Clemmensen, J., Larsen, E., and Rasmussen, M. (2012). *Forsvaret i Arktis – Suverænitet, Samarbejde og Sikkerhed*. Copenhagen: Center for Military Studies.

Rasmussen, M. (2013). *Greenland Geopolitics: Globalisation and Geopolitics in the New North*. Copenhagen: Committee for Greenlandic Mineral Resources to the Benefit of Society.

Søvndal, V. (2013). Redegørelse af 10/10 13 om arktisk samarbejde. Speech given at the Danish Parliament on 10 October 2013. Available at: http://www.folketingstidende.dk/RIpdf/samling/20131/redegoerelse/R3/20131_R3.pdf. [Accessed 4 November 2015].

5 Lightning rod

US, Greenlandic and Danish relations in the shadow of postcolonial reputations

Mikkel Runge Olesen

The triangular relations between Greenland, Denmark and the US have undergone significant transformations during the last 15 years. Global warming has opened up the Arctic in general and brought the question of resource extraction as a new revenue source for Greenland to the forefront. US military interests in Greenland, in decline during the1990s, have now re-emerged, not least after the update of the radar station in Thule to be part of the American missile defence system (Kristensen 2005). And finally, Greenland has sought and gained more autonomy from Denmark resulting, not least, in the transitioning from home rule to self-government in 2009, allowing Greenland to slowly "take home" areas of responsibility previously held by the government of Denmark (Government of Denmark 2009). Greenland's relations with Denmark and the US have never been trouble-free, however. Scandals concerning the American presence in Greenland, especially in Thule, have been a recurring feature since the late 1980s. Sometimes these scandals have been rooted in Cold War history, such as the presence of nuclear weapons in Greenland with Denmark's secret acceptance (Danish Foreign Policy Institute 1997, 185–96, 277-302, 451–84; Kristensen 2005, 185–86). Sometimes they have been related to newer US foreign policy such as alleged US rendition flight through Greenlandic airspace in the 2000s (Heiberg 2012), or the (as of February 2017) still ongoing controversy regarding the Thule Air Base service contract (KNR 2016). Common for all of these scandals, however, is that they can be blamed on the US, Denmark or both.

This chapter looks into how Greenland, Denmark, and the US's reputations shape their mutual relations. The chapter argues that the ghost of postcolonialism makes Denmark especially vulnerable in this regard by making Denmark a lightning rod for Greenlandic grievances. This is bad news for Denmark, but good news for the US. This state of affairs is anchored in both instrumental and perceptional elements. In instrumental terms, as argued by Kristensen (2005, 188), the past has, at times, been leveraged by Greenland through a politics of embarrassment, whereby Greenland has been able to use references to previous embarrassing controversies as a means of shaming concessions out of Denmark. The main focus here, however, lies on the deeply ingrained perceptional impact of the postcolonial past. The chapter, thus, argues that the reputations that each side has with each of the others, directly affect how each of them view and interpret which options, they believe are available to them in triangular relations.

The analytical approach

Traditional approaches to Greenlandic–US–Danish relations (or to each of the three bilateral relationships) have typically focused on analysing the clash of interests between the three, drawing on, implicitly or explicitly, the rationalist paradigms of International Relations (IR) (e.g. Petersen 1988; 1992; 2011; Archer 2003; Kristensen 2005; Rahbek-Clemmensen 2014). The present analysis will adopt a similar starting point. That means acknowledging that the US presence in Greenland was originally driven by military strategic concerns (Lidegaard 1996, 179–88) and acknowledging that fiscal considerations make up a core component of current Danish–Greenlandic relations. However, not least because of the historical scandals as well as the lingering presence of postcolonialism, this chapter argues for the need to also consider how previous quarrels and cooperation may affect how these nations expect each other to behave in the future and, thereby, how such perceptions may shape their attitudes to future bilateral and trilateral issues between them. How far can a partner be trusted in a given area of cooperation? In order to capture these complex dynamics, this chapter will focus on the analytical concept of "reputations" (Mercer 1996; Crescenzi 2007; Tomz 2007), understood as the entire package about what a state (or a similar other social unit) thinks of a certain partner, including both cognitive biases and emotional likes and dislikes.

Based on this definition, how can we then expect reputation to form, evolve and, above all, affect relations between countries? Within social psychology (Fiske and Taylor 1984; Fiske 1986; Leyens et al. 1994; Bodenhausen and Macrae 2000; Macrae and Bodenhausen 2001), a key finding is that we use schemata, such as the reputation we assign to others, as a cognitive processing tool to inform us regarding what to expect from others in the future. On the emotional level we can expect a similar mechanism. Transgressions done by close friends are more likely to be explained away and excused than transgressions done by outsiders (Mercer 2010, 9). Finally, it is not only the reputation that others hold of our state or nation that affects foreign policy. A nation's self-perceived reputation can also affect foreign policy. Thus, outside attacks on a nation's status as a "good" member of international society can destabilise that nation's self-esteem and trigger cognitive dissonance and shame (Finnemore and Sikkink 1998, 903–904; Risse and Sikkink 1999, 8).[1]

Social psychology also tells us that reputations are hard to change, since the mind has a tendency to reject stimuli that do not match pre-existing beliefs. It is therefore likely to happen either very slowly over time or as the result of shocks from widely disconfirming evidence (Bennett 1999, 81–85; Mercer 2010, 25). Add to this that most decision makers on both sides of a relationship between different actors generally like to take credit for successes and hand out blame for failures (Mercer 1996, 61–62), and it follows that reputations are not only hard to change in general, but also that it might be even harder to get rid of a bad reputation than it is to destroy a good reputation. Taking interests and reputation as a starting point, this chapter will use these concepts to discuss Greenlandic–Danish–American relations since the beginning of the twenty-first century.

Setting the scene: conditions and interests

The Arctic is undergoing rapid change, which is affecting the interests of all states and entities with a presence there. Before proceeding to the relationships between the Greenland, Denmark and the US, a quick outline of what is at stake for each of them is in order.

Greenland

Though there are many different views in Greenland as to *how* independence is to be achieved, the general discourse that Greenland *ought* to be an independent state is strong (Gad 2014, 4–5). On this basis, it is not surprising that a 74 per cent majority voted for Greenlandic self-government ("selvstyre") in 2008. Economic considerations remain a key obstacle, however, as Greenland is currently very dependent on a yearly Danish block grant, which, together with EU grants, amounted to 4.7 billion DKK in 2014, estimated to about 35 per cent of Greenlandic GDP (Economic Council of Greenland 2014).

Though important, the economic impact of the US's presence in Greenland is significantly smaller. Currently, the chief US contribution remains the Thule Air Base, which provides indirect revenue for Greenland through taxation of Danish and Greenlandic employees and, until recently, through profits from the service firm Greenlandic Contractors, which is partially owned by the government of Greenland. According to internal Greenlandic deliberations, total revenue amounted to DKK 159 million in 2011 (Spierman 2016, 77) about 1 per cent of Greenland's GDP. Parts of this income, especially the roughly 30 million stemming from direct profits from ownership the company (Spierman 2016, 77), has recently been jeopardised, when the US Air Force chose to replace Greenland Contractors with an American competitor. Greenlandic hopes for the Joint Committee, a committee consisting of Greenland, Denmark and the US set down to coordinate, among other things, deeper economic cooperation between the three, has yet to result in a major economic gain.

Finally, to make matters worse, Greenland is also facing challenges associated with an aging population putting further strain on Greenlandic budgets even if the block grant were to subsist at current levels (Economic Council of Greenland 2015). Self-sufficiency from the Danish block grant therefore requires either draconic savings on the Greenlandic budgets or a major new source of income.

Denmark

Denmark's chief priorities in Greenland are dual: To represent Greenlandic interests in coordination with the Greenlanders (also thereby strive to avoid further scandals), and to continue to benefit from the boost to the kingdom's influence in international affairs stemming from its Arctic status (Taksøe-Jensen 2016, 35). During the Cold War this was primarily vis-à-vis the US, but especially since the intensifying focus on the region following from global warming, the list of countries with an interest in the Arctic has widened substantially.

The US

Keeping the Thule Air Base open remains the chief priority for the Americans in Greenland. In the pursuit of this overarching regional goal comes the additional global interest in doing so without having to pay direct base rent – not so much because the US could not afford this in Thule, but because paying base rent to the Greenlanders in Thule might open up a dangerous precedence for other US bases around the world (Kristensen 2005, 195). US interests in Greenland are not limited to Thule, however. Also, in a more general sense, the US has an interest in keeping Greenland as a member of, an American-led, global community and to ensure that American businesses are well-positioned to move into Greenland if and when it should prove profitable (interview with then-US Ambassador to Denmark, Rufus Gifford, August 2016).

* * *

Basic interests outlined, the chapter will now proceed to discuss the three bilateral relationships in turn, with the aim of investigating how reputation may help us understand how each of the three entities choose to pursue their interests vis-à-vis each other.

Greenlandic–Danish relations

Former Premier Kuupik Kleist Inuit Ataqatigiit (IA) describes the historical scandals as painful "… though a lot of water has run in the creek since then …" (interview with Kleist, January 2016).[2] He also remarks, however, that they have also served as reasons for rethinking the Danish–Greenlandic relationship meaning that some good has come of them as well (interview with Kleist, January 2016). For him the problems between Denmark and Greenland today, however, have less to do with Danish ill-will towards Greenland, but instead with Danish lack of attention for Greenlandic issues resulting in Denmark forgetting to include Greenland in the decision-making process on common issues (interview with Kleist, January 2016).

Compared to Kuupik Kleist, former Premier Aleqa Hammond (Siumut) represents a much more controversial voice in Greenlandic politics marked by even greater scepticism towards Denmark. This is evident, not least, from her comments on the controversy about alleged CIA flights in Greenland which broke in 2008. Here Aleqa Hammond, then Greenlandic Minister for Foreign Affairs, remarked, likely as a pointed comment to the historical scandals, that Greenland "…unfortunately does not have good experiences regarding Danish will to represent Greenlandic interests against the USA" (Aagaard 2008).

She was not alone with this view. When the scandal broke again in 2011 as a result of Wikileaks raising suspicions regarding the Danish sincerity during the 2008 internal investigation conducted by the government of Denmark, it led former Greenlandic premier (1991–97), Lars-Emil Johansen from Siumut to voice similar concerns questioning whether Denmark could be trusted to conduct Greenlandic foreign and security policy at all (Dahlin 2011). Furthermore, similar views have also been expressed by Greenlandic Minister of Foreign Affairs, Vittus Qujaukitsoq, in an interview with the Danish newspaper *Politiken* in December 2016 (Hannestad 2016).

For Hammond, a driving factor in her mistrust of Denmark is the idea that Denmark gets more than is presently known, perhaps in the shape of discounts in NATO, for letting the US keep its base in Greenland:

> We don't know what Denmark gets out of it. It has never been told. It has been a natural thought in Greenland that of course [Greenland should] benefit from the American presence. The average citizen's understanding is that there is a man in my house, and he does not pay rent *to us* [my emphasis].
>
> (Interview with Hammond, January 2016)

Such suspicions are relatively common in Greenland (see also Henriksen and Rahbek-Clemmensen 2017). Aaja Larsen, IA MP of the Danish Parliament, shares her experiences from the 2015 election campaigns regarding the presence of this story in Greenlandic politics:

> Some of what I hear during election campaigns is also this old myth that Denmark gets a considerable discount in NATO membership because of the Thule Air Base. This, I am not able to either confirm or deny, but it is in any event something that is very present.
>
> (Interview with Larsen, February 2016)

Shifting to the Danish perspective on Greenland, this colonial legacy inevitably plays a crucial role for both Danish self-image and for Greenlandic reputation in Denmark. The Danish view of Greenland is still paternalistic (or maternalistic as Gad (2008) puts it) and is based on the notion that Greenland needs Danish help to mature as an independent state (though the 2009 Self-Government Act does give Greenland the right to decide when to pursue full independence). Denmark still perceives itself as a benevolent guardian of Greenland, which is still not ready to take care of itself (Gad 2008, 117–120). Based on this interpretation it is also much easier to understand why Danish Prime Minister Helle Thorning-Schmidt in August 2013 would reject participating in the Greenlandic-proposed commission on Danish–Greenlandic reconciliation with the words "We do not need reconciliation, but we fully respect that it is a discussion that interests the Greenlandic people" (KNR 2013). According to Heinrich (2014), this attitude may be linked to a general Danish perception of Denmark as the "good" colonial power that has nothing to apologise for. This kind of reputation, can help explain not only why Denmark has historically been very vulnerable to Greenlandic shaming based on the historical scandals, because they destabilise this story (Kristensen 2005, 118; Christensen and Kristensen 2009, 125), but also why Denmark may sometime forget to pay attention Greenlandic views.

How do these reputations affect bilateral Greenlandic–Danish relations? For the Greenlanders, the differences internally on how to view Denmark strongly affect their views on the time horizon for independence. If Denmark is currently reaping the benefits of the US presence in Thule (as Hammond suggests), it also follows that this potentially represents an alternative source of income for Greenland, which may make independence viable. For Denmark, the tendency to still see itself as the guardian that knows best is unlikely to change drastically in the short term given the resilience of reputation in general. It is therefore likely to continue to cause irritation in Greenlandic–Danish relations for the foreseeable future.

Greenlandic–American relations

Shifting focus to Greenlandic–American relations, it becomes essential to once again return to how the Greenlanders interpret two main series of historical events. First, the string of scandals from the Cold War and, more recently, the CIA's alleged rendition flights in Greenlandic airspace. Second, the continuous lack of US investments in Greenland, the disappointing track record of Joint Committee, and the ongoing crisis about Greenlandic Contractors.

Kleist talks about "… healthy mistrust …" (interview with Kleist, January 2016) of the Americans because of the scandals: "We have had so many controversial incidents surrounding the American presence in Greenland that it would be strange if we had blind trust in the Americans" (interview with Kleist, January 2016). Echoing this general distrust, Aaja Larsen argues that a reason for many Greenlanders' distrust of the US is likely rooted in the scandals because "… relations between Greenland and the US don't have much else to be based on and therefore takes up a bigger role in the relationship …" (interview with Larsen, February 2016).

The scandals can hardly be considered a new factor, however. More importantly for the recent development in American reputation in Greenland may therefore be the experiences concerning economic issues in recent years. A few years ago, when the extractive industries had their eyes on Greenland, there was the feeling in Greenland, Kleist argues, "… that the US would be able to contribute to the development of significant [economic] sectors in Greenland. And that has not happened" (interview with Kleist, January 2016). Now, in 2016, he is much less optimistic both about cooperation in the Joint Committee and about the US in general: "… [I]n relation to concrete cooperation in education or American investments or you name it, it hardly exists at all" (interview with Kleist, January 2016). This leads Kleist to conclude that "… the Americans don't take up much room anymore. The Americans don't take up much room in Greenland's consciousness" (interview with Kleist, January 2016).

Aleqa Hammond's interpretation of American reputation in Greenland differs quite markedly from Kleist's on a number of additional key issues. Asked about her view on the scandals and how they affect how she sees the US as a presence in Greenland she answers:

It is not so much about how we see the Americans. It is more about how we see Denmark. The Americans don't do anything in Greenland without official Danish consent because Greenland is too important for the Americans to let it be lost because of actions not coordinated ... [between] Denmark and the US. Therefore it is a Danish, *purely Danish*, responsibility what goes on in Greenland because Denmark, on behalf of the Kingdom, has sanctioned the American presence in Greenland.

(Interview with Hammond, January 2016, my italics)

Compared to Kleist, who generally blames both Denmark and the US for the historical scandals, Hammond's interpretation largely clears the US and places the full blame on Denmark.

Like Kleist, Hammond is quick to highlight the meagre results of the Joint Committee (interview with Hammond, January 2016). However, also here they differ in how to lay blame. Where Kleist tends to see this primarily as a consequence of American bureaucracy, Hammond also sees Denmark's presence in Joint Committee as a big part of the problem, and she is confident that Greenland can strike much better deals with the US on its own. Again, because Denmark is generally assigned more blame, as with the scandals, US reputation is in much better shape in Hammond's eyes than in Kleist's. These differences may well explain why Hammond today is significantly more optimistic than Kleist about future US–Greenlandic cooperation.

US interests in Greenland and Greenland's reputation in the US have undergone some development since the turn of the millennium. In 2003–2004, US interests in Greenland had primarily been associated with the planned upgrade of radar station at Thule. From around 2007–2008 it was clear, however, that Greenland was beginning to attract much more American attention for commercial reasons (Wikileaks 2007; 2008). The American attitude to the prospects of full Greenlandic independence, however, was initially quite sceptical. The US embassy wrote home from Copenhagen in 2006 that

Should Greenland ever strike oil and achieve independence, the United States would have as a host nation for Thule a country inclined to be sympathetic to NAM [Non-Aligned Movement] positions rather than one of our staunchest allies [Denmark].

(Wikileaks 2006)

The Non-Aligned Movement, originally an organisation for countries refraining from picking sides in the Cold War, is clearly not meant as a positive label in this respect. This outlines the relatively poor state of Greenlandic reputation in the US as the debates about independence started out. The immediate effect of this in US policy, however, seems to be relatively limited. Though initially sceptical, the US readily began to prepare for the eventuality that Greenland might someday become independent. This is seen, not least, in the attempts of the US embassy in Copenhagen to secure, though in vain, the establishment of a permanent presence point in Nuuk (Wikileaks 2007). Indeed, such mutual US–Greenlandic interest in cooperation continued to build during the resource-fever years in the end of the 2000s and in the early 2010s.

As we have seen in some of Aleqa Hammond's quotes, a key concept in the Greenlandic strategy was the idea that Greenland is geostrategically important. And indeed, this idea seems to constitute a core foundation for US views of Greenland. As former US Ambassador to Copenhagen Rufus Gifford remarked,

> because of its geopolitical significance … we have to look at Greenland and we have to look at them as an ally and a friend were they to become independent … [No matter what happens] it is going to be geographically part of North America, it's still going to be important geostrategically, so we would need to ensure that they are our loyal friend and ally.
>
> (Interview with Gifford, August 2016)

However, Gifford also stresses points of contention between the US and Greenland. According to the Ambassador Gifford the Thule Air Base service contract controversy, at least at a time, created bad blood on *all* sides (interview with Gifford, August 2016). Furthermore, along similar lines, we also find on the American side a regret that cooperation in the Joint Committee may not have proceeded as smoothly as desired in recent years (interview with Gifford, August 2016). They attribute this mainly to a mismatch of expectations especially concerning the ease with which the US could deliver economic investments to the Greenlanders remarking that

> the truth is the government can only do so much. We are talking about private investments, we are not talking about public investment and private investors make their own decisions.
>
> (Interview with Gifford, August 2016)

On this basis we can see that the Greenlandic reputation in the US today is not problem-free.

The question then becomes how important these reputations are for Greenlandic–US relations? During the period immediately following the signing of the new base agreement the Greenlandic reputational challenges in the US seem to have had only few negative consequences. This may have to do with the fact that neither Denmark nor Greenland ever attempted to involve the US in the question of independence allowing for the US to avoid choosing between the two. As expressed in a US government cable in November 2007, The US therefore felt free to focus on seizing "a unique opportunity to shape the circumstances in which an independent nation may emerge" (Wikileaks 2007). Finally, as we shall see in the next section, the limited impact of Greenlandic reputation in the US on US policy may also be attributed to the fact that Denmark at times makes use of its own reputation in the US on Greenland's behalf in order to limit the negative impact of various issues that give cause to triangular complications.

Danish–American relations

Of the three relationship pairs, the Danish–American relationship is by far the least conflictual. It is well-established in the literature that Danish reputation in Washington has been overwhelmingly positive since the turn of the millennium, not least due to Denmark's extensive military contributions to American interventions in the Middle East (Mouritzen 2007; Henriksen and Ringsmose 2011). It goes both ways. Indeed, US reputation in Danish political circles is also very good, with the notable exception of the far left. That also means that if the US supports a given policy then that carries considerable weight with many Danish politicians (Olesen and Nordby 2014, 76–81).

How do these reputations affect bilateral relations? Regarding US reputation and prestige in Denmark, it has led to the aforementioned practice in Danish politics of giving a privileged position to requests from the US. This has consequently also reinforced the favourable Danish reputation in the US. The effect of the great Danish relationship in the US, however, is a more complicated affair that can best be investigated by looking into a few instances where Denmark has *not* been, at least initially, satisfied with US policy. Greenlandic issues, especially, has often led to such situations. For instance, the case of the alleged CIA flights over Greenland was seen in Denmark as an unwanted American problem, where both Greenlandic and domestic Danish pressure on the government was high. The case was deeply problematic for Danish authorities to handle not least due to a strict US policy of "neither confirm nor deny" the existence of CIA rendition flights, and, as a consequence, the internal Danish investigation set in motion in 2008 did not manage to resolve the issue. However, not least due to Danish foreign minister Per Stig Møller's mobilisation of the great Danish reputation in the US, Denmark managed to get formal reassurances that such flights would not take place in the future without prior explicit Danish permission (Heiberg 2012: 28–37).

Another example is the current controversy regarding the Thule base service contract. Vectrus, an American company, managed to use a Danish shell company (Exelis Services A/S), to enable it to bid on the Thule Air Base maintenance contract, even though that contract is reserved for Danish or Greenlandic companies according to the 1951 US–Danish base agreement. The contract controversy has led to a series of legal battles, where Exelis first got its contract confirmed by The Government Accountability Office in February 2015, losing it in the Federal Claims Court in May 2015, and getting it reaffirmed once again in the United States Court of Appeals for the Federal Circuit in June 2016. It has also, however, seen political involvement. Former foreign minister Martin Lidegaard claims that he actively tried to use Denmark's status as a faithful ally as an argument for getting American attention on the controversy:

> We [Denmark] always deliver when needed [Lidegaard told US secretary of State John Kerry] … But now we have a situation where we need you to show that we have some goodwill. I told them that directly.
>
> (Interview with Lidegaard, April 2016)

Lidegaard's story is confirmed by interview with an American official (April 2016). This pressure led to a joint declaration by the US and Denmark in 2015 stating that:

> *The U.S. Air Force intends to conduct a new acquisition and issue a new request for proposals for the Thule Air Base Maintenance Contract* once there is agreement between the United States and the Kingdom of Denmark regarding what constitutes a Danish/Greenlandic enterprise.
>
> (United States Court of Federal Claims 2015, 19, original italics)

In 2016, foreign minister Kristian Jensen, Lidegaard's successor, brought up the issue again at a meeting with John Kerry and received Kerry's assurances that the US would look into the matter again (TV2 2016).

As the case it still ongoing, the end result is still unclear. Naturally, assurances do not guarantee results, but it is a good place to start. These episodes therefore could represent signs of a shift in dynamics from a time, especially during the Cold War, when the influence of the so-called Greenland card may have played a key role for Danish bilateral relations to the US.[3] The example above, however, shows that the flow today may now *also* run in the opposite direction, with Denmark utilising its reputation to try to affect American behaviour concerning Greenland.

Conclusion and perspectives

This chapter has shown that triangular relations between Greenland, Denmark and the US can be understood by looking at how reputations guide each of them as they pursue their individual interests. Reputations seem to have had the least dramatic influence on US policymaking. This is likely due, in part, to the fact that the US has never had to choose sides between Denmark and Greenland and has therefore been relatively unconstrained to pursue its geopolitical interests in Greenland, but it may likewise be important that Denmark has, at times, shown itself willing to use its own more favourable reputation in Washington on Greenland's behalf.

Reputations seem to have played a more important role in Danish policy formation. The neocolonial legacy has presented Denmark with serious challenges in keeping the balance between the general Danish foreign policy objective of using the Arctic to put the kingdom on the map, not least, to keep relations with the US warm, while simultaneously avoiding doing anything that might anger the Greenlanders. In this respect perceived interest in a strong bond to the US and the great prestige the US has in Danish political circles easily intermesh, while the persistent maternalistic Danish view of Greenland arguably still creates problems within the kingdom.

The most important role for reputation, however, was found on the Greenlandic side. We have seen how Denmark's very poor reputation in some Greenlandic circles has made Denmark into a sort of lightning rod for problems in triangular relations. The worse reputation Denmark has, the more of the blame for problems between the three is attributed to Denmark, and the less blame is correspondingly assigned to the US, whereas people sceptical towards the US tend to have a less bad view of Denmark. Blame, it would seem, is not infinite, but is indeed reduced when shared. For Denmark this situation is very problematic, as it goes to show, just how hard it is to repair a bad reputation. For the US, however, it means that the prospects for building a strong image in Greenland may still be quite good.

Notes

1 This also helps explain why the Greenlandic instrumental use of the past outlined in Kristensen (2005, p. 188) has often been successful.
2 With the exception of quotes drawn from American primary material or interviews with American interviewees, all quotes have been translated by the author from Danish to English.
3 The "Greenland card" has been heftily debated in Danish Cold War historiography. See Lidegaard (1996, p. 480–485) for an account of the origins of the "Greenland card" and Villaume (1997) for a critique.

References

Aagaard, C. (2008). Grønland føler sig misbrugt. *Information*. Available at: http://www.information.dk/154011. [Accessed 21 October 2015].

Archer, C. (2003). Greenland, US bases and missile defence: New two-level negotiations? *Cooperation and Conflict* 38(2), pp. 125–147.

Bennett, A. (1999). *Condemned to Repetition? The Rise, Fall, and Reprise of Soviet-Russian Military Interventionism, 1973–1996*. Cambridge, MA: MIT Press.

Bodenhausen G. and Macrae, C. (2000). Social cognition: thinking categorically about Others. *Annual Review of Psychology* 51(1), pp. 93–120.

Christensen, S. and Kristensen, K. (2009). Greenlanders displaced by the Cold War. In: M. Berg and B. Schaefer, eds., *Historical Justice in International Perspective*. Cambridge: Cambridge University Press, pp. 111–131.

Crescenzi, M. (2007). Reputation and interstate conflict. *American Journal of Political Science* 51(2), pp. 382–396.

Dahlin, U. (2011). Dobbeltspil om CIA-fly styrker ønsket om selvstændig grønlandsk udenrigspolitik. *Information*. Available at: http://www.information.dk/255922. [Accessed 21 October 2015].

Danish Foreign Policy Institute. (1997). *Grønland under den Kolde Krig: Dansk og Amerikansk Sikkerhedspolitik 1945–68*. Copenhagen: Danish Foreign Policy Institute.

Economic Council of Greenland. (2014). *Grønlands Økonomi 2014*. Nuuk: Government of Greenland. Available at: http://naalakkersuisut.gl/~/media/Nanoq/Files/Attached%20Files/Finans/DK/Oekonomisk%20raad/Gr%C3%B8nlands%20%C3%98konomi%202014%20DK.pdf [Accessed 9 March 2017].

Economic Council of Greenland. (2015). *Grønlands Økonomi 2015*. Nuuk: Government of Greenland. Available at: http://naalakkersuisut.gl/~/media/Nanoq/Files/Attached%20Files/Finans/DK/Oekonomisk%20raad/Okonomisk%20Raads%20rapport%202015.pdf [Accessed 9 March 2017].

Finnemore, M. and Sikkink, K. (1998). International norm dynamics and political change. *International Organization* 52(4), pp. 887–917.

Fiske, S. (1986). Schema-based versus piecemeal politics: A patchwork quilt, but not a blanket, of evidence. In: R. Lau and D. Sears, eds, *Political Cognition: The 19th Annual*

Carnegie Symposium on Cognition. Hillsdale, MI: Lawrence Erlbaum Associates, pp. 41–53.

Fiske, S. and Taylor, S. (1984). *Social Cognition*. Reading, MA: Addison-Wesley.

Gad, U. (2008). Når mor/barn-relationen bliver teenager: Kompatible rigsfællesskabsbilleder som (dis)integrationsteori. *Politica* 40(2), pp. 111–133.

Gad, U. (2014). Greenland: A post-Danish sovereign nation state in the making. *Cooperation and Conflict* 49(1), pp. 98–118.

Government of Denmark. (2009). Lov om Grønlands Selvstyre. Lovtidende A, no 473, 13 June.

Hannestad, A. (2016). Grønland raser over dansk "arrogance". *Politiken*. Available at: http://politiken.dk/udland/art5736298/Gr%C3%B8nland-raser-over-dansk-%C2%B Barrogance%C2%AB. [Accessed 20 January 2017].

Heiberg, M. (2012). *Et er jura at forstå, et andet land at føre. Undersøgelse af en række spørgsmål vedrørende 2008-redegørelsen om de påståede hemmelige CIA-flyvninger over og i Grønland samt dansk bistand hertil*. Copenhagen: Danish Institute for International Studies.

Heinrich, J. (2014). Forsoningskommissionen og fortiden som koloni. *Baggrund*. Available at: http://baggrund.com/une-belle-robe-demoiselle-dhonneur/. [Accessed 22 August 2016].

Henriksen, A. and Ringsmose, J. (2011). *Hvad Fik Danmark ud af det? Irak, Afghanistan og Forholdet til Washington*. Copenhagen: Danish Institute for International Studies.

Henriksen, A. and Rahbek-Clemmensen, J. (2017). *Grønlandskortet – Arktis'betydning for Danmarks indflydelse i USA*. Copenhagen: Center for Militære Studier.

KNR. (2013). Danmark på sidelinjen i forhold til forsoningskommission. Available at: http://knr.gl/da/nyheder/danmark-p%C3%A5-sidelinjen-i-forhold-til-forsoningskommission. [Accessed 28 October 2015].

KNR. (2016). Overblik: Sagen om den mistede Pituffik-servicekontrakt. Available at: http://knr.gl/da/nyheder/sagen-om-den-mistede-pituffik-servicekontrakt. [Accessed 22 August 2016].

Kristensen, K. (2005). Negotiating base rights for missile defence: The case of Thule Air Base in Greenland. In: S. Rynning, K. Kristensen and B. Heurlin, eds, *Missile Defence: International, Regional and National Implications*. Abingdon: Routledge, pp. 183–207.

Leyens, J., Yzerbyt, V. and Schadron, G. (1994). *Stereotypes and Social Cognition*. Thousand Oaks, CA: Sage Publications.

Lidegaard, B. (1996). *I Kongens Navn: Henrik Kauffmann i Dansk Diplomati, 1919–1958*. Copenhagen: Samleren.

Macrae, C. and Bodenhausen, G. (2001). Social cognition: Categorical person perception. *British Journal of Psychology* 92(1), pp. 239–255.

Mercer, J. (1996). *Reputation and International Politics*. Ithaca, NY: Cornell University Press.

Mercer, J. (2010). Emotional beliefs. *International Organization* 64(1), pp. 1–31.

Mouritzen, H. (2007). Denmark's super Atlanticism. *Journal of Transatlantic Studies* 5(2), pp. 155–167.

Olesen, M. and Nordby, J. (2014). The Middle Eastern decade: Denmark and military interventions. In: H. Edström and D. Gyllenspore, eds, *Alike or Different? Scandinavian Approaches to Military Interventions*. Stockholm: Santérus Academic Press.

Petersen, N. (2011). The Arctic challenge to Danish foreign and security policy. In: J. Kraska, ed., *Arctic Security in an Age of Climate Change*. Cambridge: Cambridge University Press, pp. 145–165.

Petersen, N. (1992). *Grønland i global sikkerhedspolitik.* Copenhagen: SNU.

Petersen, N. (1988). Grønland i Dansk sikkerhedspolitik. In: N. Petersen and C. Thune, eds, *Dansk Udenrigspolitisk Årbog 1987.* Copenhagen: Danish Foreign Policy Institute, pp. 30–51.

Rahbek-Clemmensen, J. (2014). "Arctic-vism" in practice: The challenges facing Denmark's political-military strategy in the High North. *Arctic Yearbook* 3, pp. 399–414.

Risse, T. and Sikkink, K. (1999). The socialization of international human rights norms into domestic practices: Introduction. In: T. Risse, S. Ropp, K. Sikkink, eds, *The Socialization of International Human Rights Norms into Domestic Practices.* Cambridge: Cambridge University Press, pp. 1–38.

Spierman, O. (2016). *Redegørelse om Grønlands Selvstyres deltagelse i behandling af sagen om udbud af basekontrakten vedrørende Pituffik.* Copenhagen: Bruun & Hjejle.

Taksøe-Jensen, P. (2016). *Dansk diplomati og forsvar i en brydningstid Vejen frem for Danmarks interesser og værdier mod 2030.* Copenhagen: Ministry of Foreign Affairs.

Tomz, M. (2007). *Reputation and International Cooperation: Sovereign Debt across Three Centuries.* Princeton, NJ: Princeton University Press.

TV2. (2016). Minister: Kerry erkender USA-ansvar i strid om Thulebasen. Available at: http://nyheder.tv2.dk/udland/2016-06-18-minister-kerry-erkender-usa-ansvar-i-strid-om-thulebasen. [Accessed 20 January 2017].

United States Court of Federal Claims. (2015). Consolidated post-award bid protests; limitation on competition due to international agreement; 10 U.S.C. § 2304(c); FAR § 6.302.4; inapplicability of bar on jurisdiction over actions based on treaty; 28 U.S.C. § 1502; mistake in a critical eligibility criterion; latent defect discovered by procuring agency prior to award but not corrected; considerations affecting equitable relief. Available at: http://www.wifcon.com/cofc/15-215.pdf [accessed 17 August 2016].

Villaume, P. (1995). *Allieret med Forbehold – Danmark, Nato og Den Kolde Krig.* Copenhagen: Forlaget Eirene.

Villaume, P. (1997). Henrik Kauffmann, den kolde krig og de falske toner. *Historisk Tidsskrift* 97(2), pp. 491–511.

Wikileaks. (2006). Greenland: Amb.'s visit reinforces stronger ties. Available at: https://search.wikileaks.org/plusd/cables/06COPENHAGEN705_a.html.   [Accessed   1 November 2015].

Wikileaks. (2007). Shaping Greenland's future. Available at: https://search.wikileaks.org/plusd/cables/07COPENHAGEN1010_a.html. [Accessed 1 November 2015].

Wikileaks. (2008). Your participation in the Arctic Ocean conference in Greenland. Available at: https://search.wikileaks.org/plusd/cables/08COPENHAGEN288_a.html. [Accessed 1 November 2015].

6 Chinese investments in Greenland

Promises and risks as seen from Nuuk, Copenhagen and Beijing

Camilla T. N. Sørensen

Developments in 2016 have again placed China's role and interests in Greenland in the middle of discussions and tensions between Nuuk and Copenhagen concerning the future of the Kingdom of Denmark. This chapter examines the clash of expectations, interests and concerns in relations between Greenland, Denmark and China. It shows the different, and increasingly conflicting, assessments developing in Nuuk, Copenhagen and Beijing of promises and risks associated with Chinese investments in Greenland. The chapter argues that such developments have to be analysed in the broader context of growing Greenlandic ambitions – and efforts – to resist Denmark's emphasis on the Kingdom of Denmark as one unitary foreign policy actor.

Potential large-scale Chinese investments have for a number of years played a prominent role in plans developed in Greenland and Denmark on how to foster a sustainable Greenlandic economy that with time could help ensure Greenlandic independence. Chinese companies, in particular large state-owned enterprises (SOEs) within the resource and energy sector, are highlighted on the Greenlandic side as prospective partners and investors having the necessary financial resources and the relevant skills and experiences (Kristensen 2015). As Greenland's deputy foreign minister Kai Holst Andersen put it when he visited the Polar Research Institute China (PRIC) in Shanghai in February 2014: "We particularly welcome investments from China because we can see that you can do a lot of what we need" (*China Economic Review* 2014).

The Danish government on its side acknowledges the potential benefits for Denmark in supporting a Chinese role in the Arctic region, including in Greenland, and in engaging China on Arctic issues. Denmark also played a supportive role, when China was granted observer status in the Arctic Council in 2013. However, there are growing concerns in Copenhagen about whether Nuuk is capable of dealing with large-scale Chinese investments and the potential risks that follow from such investments. These risks include an increased Chinese presence in Greenland and the political interests and pressures that may follow from such presence. Furthermore, as the geostrategic focus on the Arctic region increases, it becomes more important for Copenhagen to be able to control developments in and around Greenland that are assessed as having direct implications for Danish national security. Greenlandic politicians and public officials on their side are increasingly frustrated with what they see as signs of a Danish condescending arrogance and Danish double standards, i.e. Denmark seeks to attract Chinese investments to Denmark while expressing concerns when Greenland seeks to do the same. Nuuk therefore prefers a more intensified focus on identifying and developing common Sino–Greenlandic interests, especially within resource exploration and extraction, without too much disturbance and involvement from Copenhagen.

China has during the recent decade increased its focus on and its engagement in the Arctic. Key in China's Arctic-policy is to establish strong and comprehensive bilateral relationships with all the Arctic states and stakeholders, including Greenland, and gradually increase China's presence and influence in Arctic multilateral institutions. It is done in a careful way in order to avoid any action that would generate a fear of an assertive Chinese role – the "China threat fear" – anywhere in the world including in the Arctic region. Furthermore, it is important for Beijing not to jeopardise its strong political relationship with Copenhagen – the comprehensive strategic partnership – in place since 2008. Chinese public officials and diplomats find it difficult to understand and manoeuvre in relation to the changing legal framework within the Kingdom of Denmark and are attentive not to get dragged into what is regarded as the internal affairs of Denmark. This was, seen from Beijing, precisely what happened in 2012–13, when China's role and interests in Greenland became the focus of a heated political controversy in Denmark on whether Nuuk or Copenhagen has the authority when it comes to the resource and energy sector in Greenland and Greenland's immigration policy.

There are, however, newer initiatives and cooperation developing in recent years between Greenland and China. A memorandum of understanding (MoU) on scientific cooperation was signed in 2016, and within tourism, the "Visit Greenland" promotion activities in China have started to pay off with an increase in the number of Chinese tourists going to Greenland. Within mining, the first big Chinese investment was made in September 2016, when the Chinese company Shenghe Resources – an internationally leading company in all parts of the rare earth industrial chain – acquired a 12.5 per cent stake in the Australian-based company Greenland Minerals and Energy (GME). GME has the licence to the Kvanefjeld (Kuannersiut) mining project in southern Greenland assessed to hold major rare earth deposits and also substantial uranium and zinc deposits. The two companies have established a strategic partnership to further develop the project and there is optimism in both GME headquarters in Nuuk and in the Greenlandic Ministry of Mineral Resources that the project will move ahead.

There is a lot of complex diplomacy involved in the triangular relations between Nuuk, Copenhagen and Beijing and add to that the US component. The US is Denmark's closest strategic ally with long-standing security interests and presence in Greenland, e.g. the in Thule Air Base (Pituffik) in northwestern Greenland. How China – increasingly assessed in the US as its biggest great power rival – increases its presence and influence on Greenland is undoubtedly followed closely in Washington and is also likely an issue on the agenda in meetings and exchanges between the relevant Danish and American security and intelligence service agencies. The opening up of the Arctic and the growing presence and involvement of non-Arctic states and the evolving role of Greenland itself is therefore challenging and gradually changing the internal dynamics of the Kingdom of Denmark.

The chapter presents its analysis and argument in three steps. The first section examines and discusses recent developments in the Chinese role and interests in the Arctic region and specifically in Greenland in more detail. The second section places the issues of "China in the Arctic" and "China in Greenland" in the context of ongoing developments in Sino–Danish relations and further scrutinises the Danish and Greenlandic interests in and their assessments of promises and risks associated with Chinese investments in Greenland. The third and final section puts the parts together and discusses the clash of expectations, interests and concerns in relations between Greenland, Denmark and China. The analyses draw on interviews and meetings in Copenhagen, Nuuk, Beijing and Shanghai with Danish, Greenlandic and Chinese scholars, public officials, diplomats and businesses and on statements from Danish, Greenlandic and Chinese governments and politicians.

China in the Arctic and in Greenland – the development in the Chinese role and interests

Even though Arctic issues are not at the top of the Chinese foreign and security policy agenda, Beijing wants to be involved in the development of Arctic affairs – to be an "Arctic stakeholder" ("beiji liyi xiangguan zhe"). At the Arctic Circle meeting in Iceland in October 2015, the Chinese Foreign Minister Wang Yi (2015) in a video message described China as a "near-Arctic state" ("jin beiji guojia") and referred to China's long history of Arctic interests at least as far back as China's signing on to the Spitsbergen (Svalbard) Treaty in 1925.[1] His aim doing this was to highlight as well as legitimise the growing Chinese focus on and engagement in the region.

Overall, China's Arctic policy consists of four driving factors (cf. e.g. Zhang 2015). First, China aims to build a solid Chinese polar research capacity, which especially relates to how the melting ice and changing climate in the Arctic continuously affect China. Second, China wishes to get access to the energy and mineral resources that the Arctic holds hereby helping to secure and diversify China's energy supply. Third, China seeks to develop and get access to the Arctic sea routes, which could give China alternatives to the longer and strategically vulnerable routes in use now especially through the Malacca Strait. And fourth, China wants to be a player in the evolving regional Arctic governance regime.

Scientific interests

The core of China's activities in the Arctic region so far has been Chinese scientific interests, which include studies in geography, climatology, geology, glaciology and oceanography (Lanteigne 2015, 150). Beijing has since 2004 had its own research station in the Arctic, the Yellow River Station (Huanghe zhan) on Svalbard operated by the Chinese Arctic and Antarctic Administration (CAA) (Zhang 2015, 235–242). In 1993 China bought from Ukraine an icebreaker, now called *"Snow Dragon"* (*"Xuelong"*), which has since been on several Arctic and Antarctic expeditions and has become a symbol of China's scientific interests in the polar regions (Lanteigne 2014, 13).

In recent years the Chinese research activities in the Arctic – and in the Antarctic – have been further strengthened launching more expeditions and intensifying efforts to build networks and cooperation. A part of the explanation for this is that China, like other non-Arctic states, is taking an active part in the general science diplomacy in the region using their research activities and institutions to legitimise their overall growing presence and to strengthen their influence in the region.

In relation to Greenland, the Chinese emphasis in recent years has also been on scientific interests seeking for example to initiate more concrete cooperation between the Chinese Academy of Geological Science and the Geological Survey of Denmark and Greenland (GEUS). Indicative is also the visit to Greenlandic research institutions in September 2015 of a Chinese delegation from the Polar Research Institute China (PRIC) and the Chinese State Oceanic Administration. During the visit, the Chinese highlighted the potential for building networks and cooperation and apparently announced that the *"Snow Dragon"* is to pay a visit to Greenland in the near future (Dollerup-Scheibe 2015).[2] A further step was taken in May 2016 during yet another visit to Greenland of a Chinese delegation from the Polar Research Institute China (PRIC) and the Chinese State Oceanic Administration, when the Greenlandic Minister for Education, Culture, Research and Church Nivi Olsen and the Deputy Administrator of the Chinese State Oceanic Administration Chen Lianzeng signed a memorandum of understanding (MoU) aimed at establishing closer cooperation between Greenland and China within Arctic research specifically mentioning the establishment of a Chinese research station in Greenland as well as the exchange of researchers and students (Petersen 2016). The Chinese side is keen on moving ahead and specifically points to the Aurora Observatory in the sparsely populated Icelandic region of Kárhóll funded by the Polar Research Institute China (PRIC) as a potential model to be replicated in Greenland. The Aurora Observatory aims to further the scientific understanding on solar-terrestrial interaction and space weather (Raspotnik 2016).

Energy and mineral resources

An important driving factor is the quest for Arctic energy and mineral resources to help secure and diversify China's supply. China is operating on a long-term basis constantly working to ensure the best terms and necessary connections. Beijing therefore continuously seeks to establish strong and comprehensive bilateral relationships with all the Arctic states and stakeholders not only through the research and science cooperation mentioned above but also through economic deals and political partnerships. China has especially prioritised Iceland, indicated by the number of high-level Chinese delegations going to Iceland in recent years, and by the conclusion of the free-trade agreement (FTA) with Iceland in 2013, which was the first Chinese free-trade agreement with a European state. Chinese scholars also often point to Norway as one of the most important states for China in the Arctic – Norway has a big Arctic territory, including Svalbard with the Chinese research station mentioned above, energy and mineral resources, a proven ability to use the Arctic sea routes with direct access to the Northeast Passage (NEP) and a high-level of relevant technological knowhow. When China and Norway after six years in the freezer, in December 2016 resumed normal diplomatic relations again, the strong potential for cooperation on polar issues was mentioned in the four-point joint statement (Government of the Kingdom of Norway 2016). By establishing strong and comprehensive bilateral relationships with all the Arctic states and taking an active involvement in multilateral institutions and developments of these in the region, China seeks to ensure the most optimal platform for pursuing its interest in Arctic energy and mineral resources.

As mentioned, China adopts a long-term perspective building the bilateral relationships. This implies that despite low world market prices and the declining growth rates in the Chinese economy, China is still closely following the developing economic opportunities in the Arctic. The Chinese approach is fairly straightforward – if the Arctic becomes more open for business and shipping, China wants to have its partnerships and its capacity in the region in place. In that equation, Greenland is a case in point. Even though China seems to have downgraded its focus and interests in Greenland's resource and energy sector, the Chinese are still following developments in and around Greenland closely and if the risk and profit assessments change or new opportunities open up, China is prepared, as the recent Shenghe investment mentioned above also shows. In other words, China takes an *active* wait-and-see approach to exploration and extraction in Greenland.

The international scrutiny and anxiety directed towards Beijing's economic and resource diplomacy in the Arctic region have made the Chinese careful due to concerns about a diplomatic backlash if Beijing is perceived as taking a too-assertive approach. The Chinese are, however, doing something. The Chinese SOEs within the resource and energy sector, especially the China National Offshore Oil Corporation (CNOOC) and the China National Petroleum Corporation (CNPC), are active in northern Canada and in northern Russia as well as in the Dreki region between Iceland and Norway (Lanteigne 2014, 19). Regarding the mining of metals and minerals, which is where the Chinese focus is when it comes to Greenland, there are several Chinese companies engaged in prospecting and conducting surveys (Danish Ministry of Defence 2016, 32-36; Pu 2012). Some of these Chinese companies have also initiated cooperation with other companies. Besides the GME-Shenghe cooperation mentioned above, also Ironbark, an Australian mining company, has signed a non-binding memorandum of understanding (MoU) with the Chinese company China Non-Ferrous Metal Mining Group (NFC) for the construction and financing of the

Citronen Zinc Project in Greenland. Furthermore, there is Chinese interest involved in the potential development of an iron mine at Isua, about 150 kilometres northeast of Nuuk. General Nice Group, a large Chinese business conglomerate, also including mining companies, in December 2014 took over London Mining Greenland and hence the exploration rights at Isua. General Nice Group has experience from mining in Russia, Australia and South Africa and has expressed an interest in continuing to explore the potential in Greenland (Hornby et al. 2015).

The Greenlandic efforts to attract Chinese investments continue with frequent high-level Greenlandic delegations going to China. For example in October 2015 the Greenlandic Minister for Finances and Resources Andreas Rene Uldum participated in the annual "China Mining" conference in Tianjin. Here he met with the Chinese Minister of Land and Resources Jiang Daming and presented the Greenlandic wish for concluding a memorandum of understanding (MoU) with the Chinese Ministry of Land and Resources (MLR) similar to the MoU concluded with the South Korean Ministry of Trade, Industry and Energy in September 2012.[3] Whether the Chinese side is willing to conclude such a MoU with Greenland remains to be seen but there is no doubt that it will require strong support and reassurances from Copenhagen. Overall, the main condition for a stronger Chinese economic commitment in Greenland remains unchanged, which is that energy and resource prices and costs for exploration and extraction will have to become more economically favourable. Acknowledging this, Greenland has increased the focus on other areas in order to attract Chinese investments and develop the Greenlandic economy, e.g. export of Greenlandic fish, seal skin and tourism.

Arctic sea routes

With the melting and receding ice, new Arctic sea routes linking Asia and Europe are becoming navigable and Chinese commercial vessels have already tested the Northeast Passage (NEP) along Russia's northern coast line. For China it is approximately 30 per cent shorter than through the Strait of Malacca and the Suez Canal, but still it is not necessarily quicker or cheaper. Again the Chinese take the long-term perspective and want to be ready to exploit new opportunities. Therefore the Chinese are testing the routes and are busy designing and building new ships that are better suited to the conditions (Gang 2012, 361; Lanteigne 2014, 26–32). In 2016, the Chinese state-owned shipping company COSCO said it plans to launch regular services through the Arctic to Europe by way of the Northeast Passage (NEP) (*The Guardian* 2016). Furthermore, China's Maritime Safety Administration has in recent years released guidelines in order to promote and help Chinese ships navigate the Arctic waters. At the release of the most recent guidelines for the Northwest Passage (NWP) in April 2016, senior official Wu Yuxiao involved in drafting the guidelines stated that "Many countries have noticed the financial and strategic value of Arctic Ocean passages. China has also paid much attention" (*China Daily* 2016).

Evolving Arctic governance regimes

Generally speaking, Beijing wants to be at the table when the rules and norms in the international system are settled. This goes for the Arctic as well. The development in the Arctic is particularly interesting for China in relation to the issue of national vs. international waters. Regulations concerning this issue will further be discussed and developed as the Arctic becomes more accessible. The new sea routes increase the importance for states of being able to influence the rules for access to and transportation in the Arctic region as well as they intensify the interests and stakes involved in the still unsolved territorial and maritime disputes in the Arctic region. China wishes to be included in this process as the legal setting and governance in the Arctic region could be of relevance for China in its own region. What especially attracts China's attention in this regard is how security and territorial disputes in the Arctic region are managed between the involved states and not internationalised.

Another Arctic governance issue that China follows closely is the issue of who is and what constitutes an Arctic stakeholder. China seems to be fine with at the moment being an observer and there is still limited Chinese involvement in the Arctic Council working groups. It is likely though that the Chinese delegation in the coming years will become more active and engaged and will seek to play a bigger role (Lanteigne 2014, 9). Therefore Beijing eventually would also like to "rise" in the hierarchy of categories of states involved in Arctic affairs or in a multi-tier network of cooperation if that possibility opens up. It is also in this context that the Chinese are developing and presenting their own categories such as the above mentioned "Arctic stakeholder" ("beiji liyi xiangguan zhe") and "near-Arctic state" ("jin beiji guojia") and why it is seen as crucial for China to establish strong and comprehensive bilateral relationships with all the Arctic states. China especially fears a situation where the tension between the US and Russia affects the Arctic region. This could potentially lead to what the Chinese call a "melon effect", where sovereignty issues come to play a stronger role and where the Arctic will be split between the Arctic states isolating non-Arctic states. Such a situation could create a more negative atmosphere in the region making Chinese activities, including China's use of Arctic sea routes, more difficult (Lanteigne 2015, 154–155). Therefore, the Chinese are seeking to "lock China in".

An important part of China's Arctic diplomacy also takes place within the Arctic Circle meetings, which are highly prioritised by China. The Arctic Circle meetings established and driven in cooperation with Iceland present an attractive and useful platform for China to promote its image, i.e. Chinese soft power, and interests in the Arctic as the meetings bring together commercial, scientific and governmental interests and representatives on an equal footing. China is also increasingly seeking bilateral consultations on Arctic issues with other Arctic Council observer states such as South Korea and Japan.

So far there is no Chinese Arctic strategy, i.e. White Paper. The consensus view within China has been that Beijing's visibility in the Arctic region, unlike in other parts of the world, has not developed to the point, where such a paper is necessary either for domestic or international consumption (Hough 2012, 31; Lanteigne 2015, 150, 153). However, there is a growing debate among Chinese scholars about whether this is changing (e.g. Zhang 2015, 22–23). An important argument for why it would be preferable to present a Chinese Arctic strategy is that this could help reassure international concerns and also help focus Chinese activities and coordinate among Chinese actors in the Arctic (Zhang 2015, 22–23). That Japan at the Arctic Circle meeting in Iceland in October 2015 announced the approval of Japan's first ever Arctic White Paper will likely increase the pressure on and the incentives for the Chinese leaders to release a similar paper.

Summing up, China's role and interests in the Arctic and in Greenland are growing, but China is still acting carefully and hesitantly and generally takes a pragmatic wait-and-see approach. There is currently not much pointing in the direction of China becoming a new strong partner for Greenland. However, large-scale Chinese investments in Greenland should not be written off if it is proven profitable for China, and if such investments can be done in agreement with Copenhagen and thus do not hurt Sino–Danish relations nor challenge and hereby risk bringing the "China factor" into the complex constitutional arrangement between Nuuk and Copenhagen. As indicated above, Beijing remains confused about how to approach Greenland, e.g. on what level to meet visiting Greenlandic ministers and delegations. An illustrative example is the difficult question of whether to have a Greenlandic flag on the table – the Greenlandic delegation very much prefers that, the Danes do not mind, but the Chinese side finds this difficult to believe and is extremely attentive and careful to avoid appearing to be interfering in relations between Greenland and Denmark. Chinese investments and a generally stronger Chinese role in Greenland and in the Arctic also needs to be done in a way that does not hurt, but rather support China's efforts to build an image of China as a responsible and constructive great power. It is crucial that Chinese activities and investments in the Arctic do not strengthen the "China threat" fear or the international concerns about a more assertive and aggressive China, and Beijing therefore is careful to legitimise its presence in the Arctic often referring to its scientific interests and the interests in the Arctic sea routes.

The Sino–Danish relationship and the assessments in Nuuk and Copenhagen of China's role and interests in the Arctic and in Greenland

The relationship today between Denmark and China is at an all-time high – never before have bilateral relations been as comprehensive and wide-ranging; never before have there been so many high-level visits and dialogues on a broad range of political and strategic issues as is the case today. In April 2014, Queen Margrethe II completed the biggest-ever state visit to China, accompanied by a group of Danish ministers and business executives. The state visit was followed in September 2014 by a visit to China by the Danish Prime Minister Helle Thorning-Schmidt from the Social Democratic Party. This was her second visit to China in two years. She met with both President Xi Jinping and Prime Minister Li

Keqiang, with the overall aim of setting up new goals for "the comprehensive strategic partnership" between Denmark and China. The new government in Denmark from June 2015, led by the Danish Liberal Party, has not indicated any changes in the Danish–China policy. During his first visit to China at the end of October 2015 the Danish Foreign Minister Kristian Jensen expressed a strong Danish wish to further broaden the bilateral cooperation with China. During the visit, he also signed the necessary papers making Denmark a founding member of the Chinese-led Asian Infrastructure Investment Bank (AIIB) (Danish Ministry of Foreign Affairs 2015). In addition, a large number of visits have been made by high-level Chinese party and government leaders to Denmark in recent years, the state visit by the Chinese President Hu Jintao in June 2012 being the most prominent. Since June 2012, Danish governments have met with all seven current members of the Politburo Standing Committee. A main reason for this is that Denmark has managed to analyse and to adapt to the "new China" – especially to the implications of Deng Xiaoping's economic reforms and opening-up policies in the late 1970s for developments in China and relate these to possible Danish contributions and expertise. In this way, Denmark has become interesting and valuable to Chinese leaders. This was already the case in the early rebuilding and modernisation phase in China, but even more so since 2008 when strong Danish efforts were made to export "Danish solutions". The sector cooperation focus and the "government-to-government cooperation" ("myndighedssamarbejde") are central elements behind the success, with Danish–Chinese cooperation expanding at all levels and on many policy issues, such as education, energy and climate solutions, health and welfare, public administration, legal affairs and culture (Sørensen and Delman 2016).

How does the Arctic and Greenland fit into this? Greenland looks to China for economic commitment and investments especially in order to develop its mining industry. The development of new economic activities and industries in Greenland is also mentioned as a key priority in the 2011 joint policy document between Denmark, Greenland and the Faroe Islands – "the Kingdom of Denmark's Strategy for the Arctic 2011–2020" (Danish Ministry of Foreign Affairs 2011). In the document, it is stressed how such development requires huge investments. Copenhagen has been supportive of Greenland's intensive commercial diplomacy in China and has encouraged China's engagement with Greenland and a bigger Chinese role in the Arctic even to the point of giving reassurances to the Chinese side that it is fine to deal with Greenland directly – that does not "offend" Denmark in any way. Rather, the Danish analysis is that the Arctic region opens a potential new field of cooperation between Denmark and China especially in developing joint research on Arctic issues, e.g. climate change, where there is already some Sino–Danish cooperation established. There are also common interests in developing Sino–Danish cooperation on Arctic maritime transport, as both Denmark and China are maritime nations and thus depend on stable sea routes for economic growth.

The Danish focus on China is as an emerging market, concentrating on Danish economic and commercial interests, not on the broader foreign and security policy implications of China as a rising power (Sørensen 2016). Such an approach has also characterised the Danish analysis of China in the Arctic. As a way to deal with – and trying to influence – Chinese interests and activities in the Arctic region, Denmark has supported granting China observer status in the Arctic Council. The assessment in Copenhagen is that continued strong Danish engagement with China on Arctic issues could spill over into other fields of Sino–Danish relations and that Chinese investments placed in Greenland could also benefit Denmark and Danish businesses. The focus from both Copenhagen and Nuuk in relation to attracting Chinese investments to Greenland has broadened in recent years – it is not just the resource and energy sector, where projects in any case have a long-term perspective, but also other sectors such as infrastructure and tourism.[4]

As mentioned previously, it has not been all positive, with fear or hype developing in Denmark, especially in 2012–13, about Chinese investors, companies and workers backed by the Chinese state coming to "take over" as Greenland opens more up for exploration and extraction. In Greenland today there is disappointment and frustration over the lack of Chinese economic commitment and investments. In addition, there is mistrust in Greenland about whether Copenhagen takes enough care of Greenlandic interests in meetings and negotiations with the Chinese. As put by a Greenlandic official during a conference in Denmark on the Arctic in October 2015, Greenland increasingly wants speaking rights and does not want to have to go through Copenhagen. For example, the Greenlandic efforts mentioned above to get the Chinese Ministry of Land and Resources (MLR) to sign a memorandum of understanding (MoU) directly with its Greenlandic counterpart reflects the growing Greenlandic desire to gain a stronger and more independent negotiation position. Especially Vittus Qujaukitsoq, Minister for Industry, Labour and Trade as well as head of the Department of Foreign Affairs in Nuuk, has been very vocal in his criticism of Copenhagen being too "Denmark-centric", i.e. focusing too much on protecting and promoting narrow Danish interests, in the foreign, security and defence policy of the kingdom in general and specifically in the kingdom's Arctic strategy ignoring Greenlandic interests. He has hence called for a more Greenlandic foreign policy (e.g. KNR 2016). Such growing ambitions on the Greenlandic side clash with Denmark's emphasis on Copenhagen representing the Kingdom of Denmark as one unitary foreign policy actor. This results in increased tension and awkward episodes as Nuuk and Copenhagen struggles over speaking and seating rights. The key point here is that questions related to China's role and interests in Greenland – to Beijing's bewilderment and frustration – have played a central role in such Greenlandic–Danish arm-wrestling in recent years. One instance of this "China-factor" is the clumsy process following the suddenly announced decision of the Danish government in December 2016 that it no longer wanted to sell the former Danish naval base Grønnedal in southern Greenland. The reason given was that the base – which had not been in use for years and did not figure in the comprehensive analysis from the Danish Ministry of Defence of the future tasks and activities of the Danish Defence in and around Greenland that came out earlier that year – would still be of use in Denmark's Arctic defence.

Such re-assessment was also the official explanation given to the Greenlandic government. However, there were soon leaks indicating that the main reason for why the Danish government no longer wanted to sell Grønnedal was that the large Chinese business conglomerate, General Nice Group, already active in relation to the iron mine project at Isua mentioned above had shown an interest in buying it (Brøndum 2016). The Greenlandic government was informed by the Danish government that there had been a Chinese offer, but it was not presented as an important factor playing into Copenhagen's decision. Therefore Nuuk got very upset when information reaching them through leaks in the Danish media indicated that it was mainly in order to prevent a Chinese takeover that the Danish government decided against selling the base. The Danish government apparently did not want to point to the Chinese offer as it sought to avoid harming relations with both Beijing and Nuuk (Breum 2016). This recent case have only further strengthened the Greenlandic mistrust towards Copenhagen and the Greenlandic suspicion that the Danish government does not trust Greenlandic politicians and takes decisions regarding Greenland without involving the Greenlandic government.

Copenhagen has to look to both its relations with Nuuk and the implications for the future of the Kingdom of Denmark and to Beijing, where Denmark has so many other strong, especially economic and commercial interests invested in a positive and stable relationship. The Arctic is only a small part of the overall Danish relationship with China. On the other hand, being an Arctic state also gives Denmark an extra card to play in relations with China. However, as China becomes more active in the Arctic region, relations between Denmark and China also risks becoming more challenged by Arctic issues. It might also become more challenging for Denmark to manage its relations with China in the Arctic region if great power relations – great power competition – start to influence developments in and around the Arctic region more. NATO and the US continue to be the closest strategic allies – the main strategic framework – for Denmark, and it could prove difficult for Denmark to isolate its relations with China in the Arctic region from Denmark's relations with NATO and the US. In a situation of growing strategic mistrust and rivalry between the US and China, more pressure from the US on Denmark's China policy would be the result (Sørensen 2016).

Conclusion – the clash of expectations, interests and concerns

China has during the recent decade increased its engagement in the Arctic. Especially given Greenland's potential for resource and energy development, it is likely that Greenland will continue to be a focus of Beijing's economic and science diplomacy in the Arctic region. However, China is looking to other parts of the region as well and it is still apparent that the Arctic region will not be at the forefront of Beijing's mind as there are other parts of the world, most notably Africa, Eurasia and Latin America, which continue to have a higher priority in the Chinese efforts to secure and diversify China's energy supply. On top of concerns about whether it is profitable or not, are concerns about the international scrutiny and anxiety regarding the *real* intentions behind China's economic and science diplomacy in the Arctic. Beijing is increasingly puzzled and frustrated with the way in which its activities in the Arctic including in Greenland become politicised. Several Asian states are more active in the Arctic region these years, e.g. Japan, Singapore and South Korea, and even follow a more assertive Arctic strategy than China, but the main international focus still is on China. The Chinese confusion and difficulties dealing with the complex – and changing – constitutional set-up between Greenland and Denmark furthermore add to the Chinese hesitancy.

Denmark is developing a more ambivalent view on China's role and interests in the Arctic and in Greenland. There are strong Danish interests in attracting Chinese investments to Greenland, and Copenhagen acknowledges the potential benefits for Denmark – and Danish relations with China in general – in supporting a Chinese role in the Arctic region and in engaging China on Arctic issues. The Danish overall position favours inclusiveness involving the participation of non-Arctic states in the development of the Arctic region. However, there is also a fear of China getting too much influence and too large a foothold, especially in Greenland. Concerns about the political and security implications of prioritising and promoting a Chinese role and Chinese investments in Greenland are clearly reflected in the 2016 risk assessment report from the Danish intelligence service. Here it is underlined how large-scale Chinese investments in Greenland could bring certain dependencies and vulnerabilities and furthermore increase the risk of political interference and pressures from China on Greenlandic (and Danish) authorities. This because of the close relations between Chinese companies, especially the large state-owned enterprises (SOEs) within the resource and energy sector, the Chinese political system and the Chinese intelligence service (Danish Defence Intelligence Service 2016; Danish Ministry of Defence 2016, 54).

Consequently, there are many different – and increasingly conflicting – expectations, interests and concerns involved both internally in Denmark and in relations between Denmark and Greenland regarding "China in the Arctic" and "China in Greenland". How these relations will unfold is difficult to predict as both the interests and concerns of Denmark, Greenland and China are all changing in these years and so is the scene on which their relations play out – the Arctic region. There probably will be more and more direct contacts between Nuuk and Beijing, which is what the Greenlandic side prefers as reflected in their MoU efforts and also in their annual foreign policy reports, where there is a growing demand for establishing a permanent Greenlandic representation in Beijing. Copenhagen is unlikely to openly resist these efforts being sensitive to Greenlandic criticism and attentive to avoid more awkward episodes and scandals. Nuuk increasingly seems to see the establishment of a good bilateral relationship with China as crucial for realising Greenland's aspirations for independence. However, again, the Chinese will continue to be cautious and attentive as to whether China is dragged into conflicts between Nuuk and Copenhagen. Beijing, however, seems to be keen on moving ahead on exploring the potential for scientific cooperation, so this might be the area where Greenland first experiences Chinese economic commitment.

Scientific cooperation with China, in contrast to Chinese resource exploration and extraction in Greenland, seems easier for Copenhagen and Nuuk to agree on and to manage.

Acknowledgements

The author would like to thank the many Danish, Greenlandic and Chinese scholars, public officials, diplomats and businesses, who were willing to meet and participate in interviews. The research has also benefitted from financial support from the Asia Dynamics Initiative (ADI) at the University of Copenhagen as well as the China Nordic Arctic Research Center (CNARC) fellowship program.

Notes

1 The shortest distance between China's northernmost point in Heilongjiang province and the Arctic Circle is approximately 1400 kilometres.
2 The Chinese State Oceanic Administration is under the Ministry of Land and Resources in Beijing and covers among other areas protection of ocean environment and China's coast.
3 Cf. the press statement from China Mining available at: http://www.chinaminingtj. org/en...../index.php/en-media/news/1024-2015-10-23-12 (Assessed January 11, 2016).
4 Cf. the press release from the Greenlandic government following the visit to Beijing in October 2015 of a big Greenlandic delegation led by Vittus Qujaukitsoq, Minister for Industry, Labour and Trade as well as head of the Department of Foreign Affairs in Nuuk – available at: http://naalakkersuisut.gl/da/Naalakkersuisut/ Nyheder/2015/10/3010115-Greenland-event-i-Beijing (Assessed August 16, 2015).

References

Breum, M. (2016). Løkke stopper kinesisk opkøb i Grønland. *Information.* Available at: https://www.information.dk/indland/2016/12/loekke-stopper-kinesisk-opkoeb-groenland [Accessed 17 February 2017].

Brøndum, C. (2016). Danmark forhindrer kinesisk opkøb af marinebase i Grønland. *Nordic Defence Watch.* Available at: http://www.defencewatch.dk/danmark-forhindrer-kinesisk-opkoeb-marinebase-groenland/. [Accessed 17 February 2017].

China Daily. (2016). China charting a new course for maritime transportation. 20 April. Available at: http://english.gov.cn/news/top_news/2016/04/20/content_281475331301933.htm. [Accessed 15 August 2016].

China Economic Review. (2014). Greenland, a frontier market unlike any other for China. 27 February. Available at: http://www.chinaeconomicreview.com/china-in-the-arctic-greenland-iron-mining. [Accessed 11 January 2016].

Danish Ministry of Defence (2016). *Sikkerhedspolitisk redegørelse om udviklingen i Arktis.* Available at: http://www.fmn.dk/nyheder/Documents/arktis-analyse/bilag2-Sikkerhedspolitisk-redegoerelse-om-udviklingen-i-arktis.pdf. [Accessed 16 August 2016].

Danish Ministry of Foreign Affairs. (2011). *Kingdom of Denmark Strategy for the Arctic 2011-2020.* Available at: http://um.dk/en/~/media/UM/English-site/Documents/

Politics-and-diplomacy/Greenland-and-The-Faroe-Islands/Arctic%20strategy.pdf. [Accessed 6 November 2015].

Danish Defence Intelligence Service. (2016). *Intelligence Risk Assessment 2016.* Available at: https://fe-ddis.dk/eng/Products/Documents/Intelligence%20Risk%20 Assessment%202016.pdf [Accessed 17 February 2017].

Danish Ministry of Foreign Affairs. (2015). Danish foreign minister broadens bilateral cooperation with China. Press Release. Avaliable at: http://kina.um.dk/en/news/ newsdisplaypage/?newsID=B950E101-6B9E-4167-B09E-7DB0C361A0E5. [Accessed 6 November 2015].

Dollerup-Scheibel, M. (2015). Kinesisk forskningsskib til Grønland. *Sermitsiaq,* 27 September Available at: http://sermitsiaq.ag/kinesisk-forskningsskib-groenland. [Accessed 6 November 2015].

Gang, C. (2012). China's Emerging Arctic Strategy. *Polar Journal* 2(2), pp. 358–371.

Government of the Kingdom of Norway (2016). Statement of the Government of the People's Republic of China and the Government of the Kingdom of Norway on Normalization of Bilateral Relations. Avaliable at: https://www.regjeringen. no/globalassets/departementene/ud/vedlegg/statement_kina.pdf.　[Accessed　16 February 2017].

Hornby, L., Milne, R. and Wilson, J. (2015). Chinese group General Nice takes over Greenland Mine. *Financial Times.* Available at: https://www.ft.com/ content/22842e82-9979-11e4-a3d7-00144feabdc0 [Accessed 22 May 2017].

Hough, P. (2012). *International Politics of the Arctic: Coming in from the Cold.* Abingdon: Routledge.

KNR. (2016). Vittus ønsker mere grønlandsk udenrigspolitik. Available at: http://knr. gl/da/nyheder/vittus-%C3%B8nsker-mere-gr%C3%B8nlandsk-udenrigspolitik. [Accessed 17 February 2017].

Kristensen, R. (2015). Kinesiske råstofinvesteringer i Australien og Canada – erfaringer for Danmark og Grønland. *Politica* 47(2), pp. 216–233.

Lanteigne, M. (2014). *China's Emerging Arctic Strategies: Economics and Institutions.* Reykjavik: Center for Arctic Policy Studies.

Lanteigne, M. (2015). The role of China in emerging Arctic security discourses. *Sicherheit und Frieden* 33(3), pp. 150–155.

Petersen, S. (2016). Tættere forbindelser indenfor arktisk forskning med Kina. *Sermitsiaq.* Available at: http://sermitsiaq.ag/taettere-forbindelser-indenfor-arktisk-forskning-kina. [Accessed 12 August 2016].

Pu, J. (2011). Greenland Lures China's Miners with Cold Gold. *Caixin.* Available at: http://www.caixinglobal.com/2011-12-07/101016220.html.　[Accessed　23　May 2017].

Raspotnik, A. (2016). Solar-terrestrial interaction between Iceland and China. *High North News.* Available at: http://www.highnorthnews.com/solar-terrestrial-interaction-between-iceland-and-china/. [Accessed November 22 2016].

Sørensen, C. (2016). A small state maneuvering in the changing world order: Denmark's "creative agency" approach to engagement with the BRICs. In: S. Fryba and L. Xing, eds, *Emerging Powers, Emerging Markets, Emerging Societies: Global Responses.* Basingstoke: Palgrave Macmillan, pp. 211–234.

Sørensen, C. and Delman, J. (eds) (2016). Tema: Dansk Kina-politik – Fra spørgsmål om eksport og danske arbejdspladser til ny verdensorden. *Økonomi & Politik* 89(1), pp. 2–74.

The Guardian (2016). China sets its sights on the Northwest Passage as a potential trade boon. 20 April. Available at: https://www.theguardian.com/world/2016/apr/20/china-northwest-passage-trade-route-shipping-guide. [Accessed 16 August 2016].

Wang Yi. (2015). Video message by foreign minister Wang Yi. Ministry of Foreign Affairs of the People's Republic of China. Available at: http://www.fmprc.gov.cn/mfa_eng/wjdt_665385/zyjh_665391/t1306857.shtml. [Accessed August 12 2016].

Zhang, X. (2015). *Zhongguo Beiji Quanyi yu Zhengce Yanjiu*. Shanghai: Current Affairs Publishing House.

7 The politics of economic security

Denmark, Greenland and Chinese mining investment

Kevin Foley

A heated political controversy emerged among Danish political and security elites between 2012 and 2014 regarding the security and foreign policy implications of Chinese investment in Greenland. The points of contention were diverse, and included concerns that Chinese investment in Greenland's mineral resources would increase Chinese influence in Greenland's domestic politics, allow state-owned firms from China to place personnel and infrastructure along ports associated with mining projects, and allow China to lock up supplies of critical raw materials such as rare earth elements and uranium.

Greenland has held a unique strategic value to Denmark and the United States since the early days of the Cold War. The far north of the island is host to Thule Air Base, which plays an important role in tracking satellites and intercontinental ballistic missiles. Greenland is also positioned along shipping lanes that are expected to be strategically important at some point in the future as Arctic ice continues to melt, and Greenland's potential endowment of rare earth elements is also often cited as evidence of the island's strategic value. In the context of Greenland's newfound political autonomy (self-government) from Denmark following a 2008 referendum, the prospect of increased Chinese investment on the island was troubling to many in Denmark's political elite.

This narrative, however, is undermined by several important and often overlooked facts. First, domestic entrepreneurial actors including the government of Greenland and British and Australian mining investors played an active role in marketing mining projects in Greenland to potential Chinese investors, an unsurprising strategy given the prominence of Chinese mining companies in global markets during that time period. The active role that these actors played in reaching out to Chinese investors contradicts the widely-held suspicion that China was seeking to invest in mining operations on the island as part of a strategic plan to achieve influence in Greenland or the Arctic. Second, despite such active efforts to attract Chinese capital for a major mining project, no investments were made in Greenland, nor were there any Chinese workers on the island.[1]

The company that attracted the most controversy was the London Mining Company (LMC), a small British mining company that had purchased a claim to an undeveloped iron ore mine in the Isua Fjord near Nuuk, Greenland's capital. Danish and international press reports occasionally referred to LMC as a Chinese

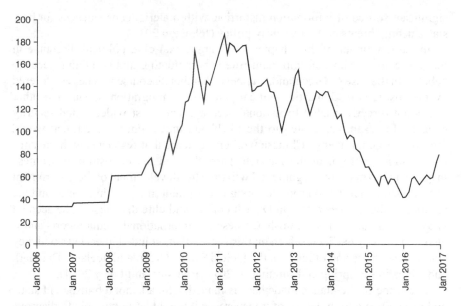

Figure 7.1 China import iron ore fines 62% FE spot price (CFR Tianjin port), US dollars per dry metric ton (2005–2015). Source: Index Mundi 2015.

mining company, but this was inaccurate. LMC had entered into talks with a group of Chinese firms to invest in the project as early as 2011, but these talks were unsuccessful and fell apart by 2013. Indeed, any plans to attract Chinese or other international investment to the Isua project seemed increasingly improbable following the collapse of iron ore and other mineral prices in late 2011 and, after a short rebound, again in 2012 (see Figure 7.1).

Although efforts to attract Chinese investment were unsuccessful, the subject of Chinese investment in Greenland was debated intensely by Danish political elites. These debates were expressed in terms of national security, and authoritative sources in Danish security circles began to seriously consider the risk of increasing Chinese influence in Greenland. For example, an evaluation of this risk has been included in the list of security risks to Denmark in the annual *Intelligence Risk Assessments* of the Danish Defence Intelligence Service (DDIS) in every year since 2011 (Danish Defence Intelligence Service 2011–16).

Chinese mining companies investing overseas are no strangers to political controversy, and proposed Chinese investments in many countries have been met with opposition organised around national security concerns. But the preceding correction to the conventional narrative suggests that political discourse can take on a life of its own, contributing in material ways to perceptions of China as a security threat in the absence of evidence suggesting that these perceptions are warranted. As such, the Greenland case has broader implications for how we think about international security and perceptions of China's rise. As Alastair Iain Johnston has argued, the "echo chamber" effect in international relations can be a significant source of information distortion, with material consequences for how states define threats and set security policy (Johnston 2013).

In the remainder of this chapter I will argue that elite political discourse in the host country plays an important role in perpetuating and sustaining security debates. In the case of Greenland, it appears that this discourse was heavily filtered by the positions of domestic political parties on immigration, labour and other issues not directly related to national security. The most widely cited security concern, for example, related to the likelihood that Chinese investment would require a large number of Chinese workers, a theme that resonates with the anti-immigration platform of the far right party that led the opposition to Chinese investment. To make this argument, I will trace the development of the government of Greenland's efforts to promote foreign investment in its mining sector and the reaction to these investments in Danish politics and elite discourse.[2] The account is based on an analysis of Danish, Chinese and international media reports and is also informed by earlier research on this subject that included interviews with key actors, academics and officials in the United States, Greenland, Iceland, Denmark and China in the spring and summer of 2014 (Boersma and Foley 2014).[3]

The scope of the claims made here is limited. China is not a transparent state, and even given a high degree of transparency it would be impossible to disprove claims that investments made by Chinese firms are part of a national security strategy. The assumption in this chapter is that such suspicions can only be reasonably evaluated based on observed behaviour. The aim of this chapter, then, is not to argue that China does or does not have any particular strategic intentions regarding Greenland. Instead, I hope to show that speculation and political rhetoric far exceeded actual developments. I conclude it would be misleading to suggest that evidence from the Greenland case can lend support to arguments about Chinese strategic interests in the region.

Background: autonomy and mining

The government of Greenland's efforts to raise capital for mining projects were closely linked to the government's need to fund political autonomy from Denmark. In November 2008, Greenlanders voted in a referendum to become politically autonomous from Denmark (self-government). This change in Greenland's status took effect in June 2009 with the passing of Greenland's Self-Government Act, and in December 2009 a new minerals law was passed that granted Greenland the right to exploit to underground mineral resources. The development of mining and offshore oil and gas projects became the clearest, and possibly only, path towards raising sufficient funds to cover anticipated budget shortfalls. For Greenland, then, there was (and remains) a powerful political incentive to quickly develop mining projects. The government of Greenland saw the rapid development of capital-intensive resource extraction projects as critical for the goal of full autonomy from Denmark, but the government's sense of urgency may have also contributed to misgivings on the Danish side over the implications of a closer relationship with China at just the moment that Greenland was pursuing greater autonomy from Denmark. This section provides an outline of the political context of Greenland's development strategy and the campaign to raise capital for mining projects.

The Self-Government Act made the government of Greenland responsible for financing the cost of all areas for which it assumed administrative responsibility and transferred ownership of relevant state assets to the government of Greenland. At the same time, it fixed the annual block grant from Denmark to Greenland at real 2009 levels, a change from the earlier system in which the transfer was periodically renegotiated as government costs increased. Greenland's economy is small and highly dependent on subsidies from Denmark, and the block grant accounts for roughly half of Greenland's government spending and 25–30 per cent of GDP (Vahl and Kleeman 2014). Greenland also faces increasing social welfare costs and lower revenue in the future due to a rapidly aging demographic profile (Economic Council of Greenland 2013).

Mining is not new to Greenland. Its mineral resources were actively explored during earlier periods of high commodity prices, and as the terms of Greenland's autonomy (self-government) were being negotiated, the exploitation of mineral resources was apparently anticipated to be sufficient to cover future budget shortfalls. The plan to finance this budget shortfall by developing mining projects was a key part of the government of Greenland's strategy from the beginning of the self-government negotiations and was central to the terms of Greenland's self-government. The new mineral resources law, for example, included an agreement to pay Denmark 50 per cent of all resource revenues beyond the first 75 million DKK (USD 11.46 million at current prices) to offset Danish support to Greenland via the block grant (Boersma and Foley 2014).

The government of Greenland may have been overly optimistic in its projections for the development of major mines on the island, but there were several resources that had already been established for years, most notably a very large iron ore deposit in the Isua Fjord. The Isua claim was first identified in 1965 and was explored by Marcona, a Peruvian iron miner, in the 1970s and again by Rio Tinto in the mid-1990s. London Mining purchased the claim from Rio Tinto in 2005 and conducted some drilling and exploration between 2008 and 2011 and marketed the project to international investors (London Mining Co. 2014).

Three major hurdles stand in the way of actually developing a working iron mine at the Isua site. The first is the relatively high cost of extracting iron ore at the Isua site given the technical challenges associated with building and operating a mine in a subarctic climate at the edge of a glacier. This problem was resolved in the short term by record prices for iron ore, which increased rapidly along with other commodity prices in the late 2000s and peaked in the first half of 2011.

The second problem is a lack of funds. London Mining estimated that it would cost USD 2.35 billion to construct the Isua mine and sought financing from international investors. As a small, primarily financial investor without the capital or resources to build and operate a large mine like the one proposed for Isua, London Mining needed an investor to purchase the mine and take over its construction and operation. The traditional approach would have been to sell the mine to a mining "major", i.e. one of a handful of large, multinational mining companies that dominate the industry such as Rio Tinto or BHP Billiton.

A review of London Mining's promotional materials and presentations to the government of Greenland makes it clear that the company planned to seek Chinese financing and engineering for the project. This is not surprising given the state of global investment markets between 2009 and 2012. Chinese mining firms began to invest overseas in large numbers only around 2004, but rapidly increased over the course of the 2000s on rising domestic demand and in line with industry trends towards greater vertical integration (Humphreys 2014).

In the wake of the global financial crisis, Chinese mining firms continued to increase their overseas investment in the mining industry while major mining companies from the rest of the world cut back. This meant that for several years after the financial crisis, China was one of the most likely sources of capital for mining projects. According to a report by PriceWaterhouseCoopers, Chinese investors accounted for 22 per cent of global mining mergers and acqusitions by 2009 (PriceWaterhouseCoopers 2009). Chinese investment was also more likely to go to relatively risky projects with higher operating costs.

The third major challenge facing the government of Greenland was the lack of skilled labour in Greenland. London Mining's plan was to attract Chinese investors to the project, and the company's presentations and reports to the government of Greenland in 2013 estimated that between 1,500 to 3,000 Chinese or other Asian workers would be needed during the construction phase of the project, and that a permanent workforce of around 800 workers would be needed to operate the mine (London Mining Co 2013). The Government of Greenland passed legislation in 2012 to facilitate London Mining's campaign to raise capital for the project. The Large Scale Projects Act (Storskalaloven) established a framework for foreign companies to contract with foreign workers on collective agreements for mineral and hydropower projects exceeding DKK 5 billion (USD 760 million in 2016), and for which local skilled labour was insufficient to meet project needs (Boersma and Foley 2014). Given Greenland's very small and geographically dispersed population and the lack of experience in the mining sector, it is understandable that the government passed the legislation. Nevertheless, the law aroused controversy in both Greenland and Denmark, and this controversy in turn heightened suspicions surrounding Chinese intentions in Greenland.

In response to this controversy, which is covered in more detail in the last section, London Mining clarified in March 2013 that the project would require a workforce that was initially 20 percent Greenlandic, 45 percent "Chinese/Asian," and 35 percent "Western." After the first four years of operation, London Mining estimated that it would have trained workers from Greenland in sufficient numbers that they would represent 55 percent of the workforce, so that Chinese or Asian workers would no longer be needed. London Mining presumably hoped that this clarification would deflect political pressure surrounding the labour issue, which would have made it difficult to raise capital from Chinese investors. At the same time, however, these precise estimates give the misleading impression that a deal had been reached with a Chinese investor. This was not the case. At the same time, the fact that a mining company felt it necessary to give what amounted to a racial breakdown of workers is a sign that by late 2012 the politics of the case had begun to take on xenophobic overtones.

In addition to its legislation in support of the Isua project, the government of Greenland actively promoted the island as a destination for mining investment. Representatives of the government of Greenland travelled throughout this period to major mining investment conferences, making annual stops at the Prospectors and Developers Association of Canada (PDAC) in Toronto and at the China Mining Congress in Tianjin, a major forum for investment deals with Chinese mining companies (Høegh 2012). This included a visit to the China Mining Congress in 2011 by Ove Karl Berthelsen, then Greenland's Minister for Industry and Mineral Resources. During this visit, discussions were apparently initiated with Sichuan Xinye, a mining company owned by the government of Sichuan province. The following year saw a reciprocal visit to Greenland in April 2012 by Chinese Minister of Land and Resources Xu Shaoshi and Liu Cigui, Director of the Chinese State Oceanic Administration, and a visit in August 2012 by a delegation organised by the China Development Bank for meetings with London Mining. By late 2013 or even as early as the end of 2012, Sichuan Xinye had withdrawn from negotiations citing technical and financial issues with the Isua project (Jun 2012; 2013). These negotiations with Xinye, which also involved other potential consortium members including the China Development Bank and a state-owned construction company, were the closest Greenland came to attracting large-scale Chinese investment.

Why did Xinye withdraw from the deal? It is possible that the political controversy in 2012 regarding the Large Scale Projects Act persuaded the company to steer clear of the project. However, it is not clear from available reports that this was a decisive factor, or even that the two companies had come close to a deal. The more straightforward explanation is simply that by 2012 and 2013, market conditions no longer justified a major mining project anywhere in the world, much less in a subarctic fjord with relatively high projected costs. A sharp decline in iron ore prices beginning in April 2012 (see Figure 7.1) likely signalled the end of any serious interest in investments in iron ore projects in Greenland for the near future. Prices of other metals that might potentially be developed into mines in Greenland, including copper and rare earth elements, have also fallen dramatically. Suffering financially from this market environment and from losses related to the Ebola crisis at the company's only operating mine (an iron ore mine in Sierra Leone), London Mining went into administration in October 2014. A privately owned Chinese investor purchased the rights to the Isua claim in January 2015 but does not appear to be capable of, or interested in, actively developing the project (Boersma and Foley 2015).

Several other Chinese mining and engineering companies have been active in Greenland and are worth mentioning briefly here. A small group of geologists from a consortium of provincially owned firms from Jiangxi Province has been conducting geological exploration activities in a remote area of northeastern Greenland, and an Australian zinc project in northern Greenland has partnered with China Nonferrous Metal Industry's Foreign Engineering and Construction Co., Ltd. (NFC) for engineering and construction. A rare earth project in southern Greenland has signed two memoranda of understanding with a Guangdong-based subsidiary of NFC for downstream separation of rare earth elements at a major new separation facility under construction in China. Unlike Sichuan Xinye, which was considering taking on the Isua project as a major investor, the other Chinese companies active in Greenland are operating at a smaller scale. The Jiangsu companies are engaged in exploration, and after identifying a major claim would seek to sell to a larger investor. NFC's current memorandum with Ironbark, the Australian company behind a zinc mine in northeast Greenland, includes an option for a minority financial stake in the project, but they are principally committed only to engineering and construction in the Ironbark zinc mine and to distribution in the Kvanefjeld rare earths project.

The presence of multiple Chinese mining companies in Greenland should not be particularly surprising given that Chinese companies have been among the most active global resource investors since the late 2000s. This new prominence of Chinese companies has been particularly notable as the largest mining majors have cut back investments in a down market. As Humphreys (2014) notes, this development is consistent with two recent trends in the mining industry: first, the industry has seen increasing vertical integration by iron and steel companies investing directly in mines. Second, mining companies from developing countries, including not only China but also Brazil, India, and Russia, have greatly increased their overseas investments in recent years. In Greenland, South Korea has also considered investments in the mining sector. In 2012 South Korean President Lee Myung-Bak visited Greenland and signed four memoranda of understanding, two of which were related to resource development (Yonhap News Agency 2012). Since that time, the Korea Resources Corporation (KORES) has been active in exploration in Greenland and at one point entered into a partnership with the now-defunct NunaMinerals A/S.

Danish concerns about Chinese intentions in the Arctic are not solely informed by the Greenland case. The argument presented in this chapter is that external developments contributed to a narrative of Chinese strategic interests in the Arctic that was highly salient to Denmark given its status as an Arctic littoral state, and that domestic political discourse sustained and amplified these suspicions. The next section traces the origins of this narrative to a puzzling Chinese real estate investment in Iceland. The last section will trace the political debates as they played out in Denmark's parliament and media.

External sources of security concerns

Danish concerns about China did not emerge out of thin air. They can be traced back to a 2011 bid by Huang Nubo, a Chinese real estate investor, to purchase land in a remote part of Iceland (BBC 2011). This proposed land deal aroused controversy in Iceland amid suspicions that the Chinese buyer harboured ulterior strategic motives, and the bid was ultimately rejected on the grounds that Huang's investment company could not be exempted from Icelandic laws preventing foreign nationals from purchasing land. In interviews with the press, Huang attributed the controversy to "internal struggles" between Iceland's political parties (Jackson and Hook 2011). Whatever the reason for the rejection of the land deal, it appears that many in Iceland found it difficult to understand why a wealthy individual from China with no ties to the Chinese state would seek to purchase a large tract of land in a remote corner of the country. Because the deal seemed to make little sense from a business standpoint, many commentators assumed that there was likely to be a hidden agenda behind the deal. Interior Minister Ogmundur Jonasson, for example, wrote that "we face the fact that a foreign tycoon wants to buy 300 sq km of Icelandic land. This has to be discussed and not swallowed without chewing" (BBC 2011). Baldur Guðlaugsson, Iceland's former Permanent Secretary of Finance, questioned in an opinion piece whether Huang's business might in fact be a cover for Chinese state interests (Guðlaugsson 2011). Ultimately, Huang was unable to persuade sceptical Icelandic lawmakers, and in particular the Minister of the Interior with the authority of approving the deal, that his plans to develop an eco-resort for wealthy Chinese tourists were in earnest.

Huang's bid attracted attention outside of Iceland, including in Denmark. In an August 2011 blog post titled "Dr. No on the Volcanos" that was reported in the Icelandic media, Former Danish Foreign Minister Uffe Ellemann-Jensen compared Huang to a James Bond villain and echoed concerns already voiced in Iceland that Huang may have been secretly motivated by geopolitical interests related to Iceland's position in the Arctic (Ellemann-Jensen 2011). In January 2012, just over a month after the high-profile rejection of Huang's bid, a security-oriented think tank in Copenhagen released a report calling for greater attention to Arctic security policy. The report, written by a former head of the Royal Danish Navy, cited close ties between China and Iceland as evidence of the new security threat posed by China (Wang 2012). The report also flagged Greenland's rich endowment of rare earth elements as a resource of potential strategic value, but it did not link this issue to China.

Although officials in Greenland appear to have been openly cultivating Chinese investment for years, the Danish debates about China and Greenland did not really begin until the days preceding the June 2012 state visit to Copenhagen by Chinese president Hu Jintao. On June 12, 2012 *Politiken*, a leading Danish newspaper, reported on London Mining's application process for the Isua mine. The article incorrectly referred to London Mining as "Chinese-controlled," ("et kinesisk kontrolleret selskab…") and quoted Rear Admiral Nils Wang, the author of the recently published Arctic Security report, on his views on the Chinese strategic thinking behind the investment. China, according to Wang, had a long-term strategic interest in the region and interests in Greenland's natural resources (Halskov and Davidsen-Nielsen 2012). The developments in Iceland appeared to prompt a reconsideration of a process that had already been in progress for many years. Officials from Greenland had travelled to China as early as 2005 and again in 2011, and the latter visit was explicitly focused on seeking mining investment (Ministry for Foreign Affairs of the People's Republic of China 2005; Ministry of Land and Resources of the People's Republic of China 2011). The idea of bringing Chinese workers to build industrial projects in Greenland was also not new, and had been discussed as early as 2008 in the context of a planned (but never completed) aluminium smelting project in the town of Maniitsoq (Nielsen 2008). What had changed was the perceived security environment in the wake of the controversy in Iceland. The political environment had also changed since 2008, with the centre-right Liberals and far right DPP parties in a parliamentary opposition bloc and a new government led by the centre-left Social Democratic Party in power since 2011.

Political debates and security concerns

As the government of Greenland began to actively promote international mining investment 2011 and 2012, a move that was prompted by the 2008 self-government referendum and the 2009 passage of the Minerals Act, debates emerged in Denmark over the legality of several unrelated aspects of legislation in Greenland related to mining. As discussed in the introduction to this chapter, those debates were framed on narrow issues related to labour, immigration, uranium mining, and securing access to critical raw materials for EU member states, but in each case the issues were also framed in terms of strategic concerns about China. The political significance of these debates should be understood in the context of Danish politics, and in particular the electoral defeat in November 2011 of the centre-right government that had been in power since 2001. This government consisted of a coalition between the centre-right Liberals (Venstre) and the Conservative People's Party (Det Konservative Folkeparti, or DKF), with support from the far-right Danish People's Party (Dansk Folkeparti, or DPP). The centre-right government that was in power throughout the long process leading up to the 2008 referendum in Greenland now found itself in the opposition, and developments in Greenland provided an opportunity for these parties to paint the ruling coalition led by the Social Democrats (Socialdemokraterne) as weak on foreign policy, immigration and labour. In addition, the terms of the 2009 Self-Government Act limit Danish intervention to the areas of foreign and security policy, including immigration, which provided a convenient legal basis for these criticisms.

It is no coincidence that opposition to Chinese mining in Greenland was framed in terms of immigration policy. The first major policy initiative of the incoming coalition led by the Social Democrats was to unwind a series of strict immigration controls and anti-immigrant policies that had been put in place by the previous government. The government abolished many restrictions put in place by the conservative coalition and introduced major substantive changes including the granting of Danish citizenship to children of immigrants born and raised in Denmark and the abolition of the Ministry for Refugees, Immigrants and Integration that had been previously established by the conservative coalition (Adams 2011).

The controversy that emerged over Greenland thus presented an opportunity for the opposition parties to combine their core domestic policy issues with a political critique couched in national security terms. As discussed above, the passage in

2012 of the Large Scale Projects Act by Greenland's parliament was intended to facilitate investment in major mines, and in particular London Mining's efforts to secure foreign investment in the Isua project. Because London Mining was reported to be in talks with a Chinese mining company to invest in the mine, criticism of the Large Scale Projects Act was closely associated with criticism of the prospect of Chinese investment in Greenland and in particular with the consequences of having thousands of Chinese workers coming to Greenland. It is worth quoting at length a passage in a November 2012 newspaper column by a centre-left commentator in opposition to the Large Scale Projects Act:

> Is it beneficial that China will fly in its own workers [at a time] when Greenland has a nasty unemployment problem? Is it beneficial if China allows its own worker ants to toil as slaves for lower wages than Greenlanders receive? Is it a beneficial that Chinese workers will not have the same bargaining rights as workers from Greenland? Is it beneficial that the Chinese will live in segregated camps and eat rice boxes and noodles instead of Greenlandic food? ... And is it in the interest of Greenland to become dependent on the world's next superpower, which has long shown that it is ice cold to the autonomy of peoples (see the occupation of Tibet, the control of Hong Kong and the threats against Taiwan), just as it is shockingly indifferent to oppression in other countries (Sudan and Myanmar)? Is China really a good partner for Greenland?
>
> (Jerichow 2012)

This passage illustrates the fierce opposition with which the idea of Chinese investment was met in some corners of the Danish commentariat. It also shows how this opposition was articulated in a way that echoed common themes in Denmark's domestic politics, in particular immigration and the protection of organised labour. And yet Greenland's proposal of a special labour law for large infrastructure projects was arguably a necessary step for the promotion of resource investment. It appears that the risk that the law would lead to immigration problems for Denmark or Greenland was minimal, particularly in light of the remote location of the proposed mining projects.

The position of the centre-left governing coalition was that Greenland's Large-Scale Act was within the bounds of the new Self-Government Act and therefore should not be opposed by the Danish government. Opposition to the Act came from the centre-right Liberal Party (Venstre), the far-right Danish People's Party (DPP) (Dansk Folkeparti) and the far-left Red-Green Alliance (Enhedslisten). In January 2013, the Liberal Party's leader and former Prime Minister, Lars Løkke Rasmussen, called for the Danish government to postpone the implementation of the Greenlandic Large-Scale Act in order to consider alternatives (Winther 2013). Former Liberal Party Finance Minister Claus Hjort Frederiksen issued a proposal for a Danish fund that would promote Nordic and European investment in Greenland's mining sector. The fund was explicitly conceived of as an alternative measure that would prevent non-European investment in Greenland's mining sector, and was endorsed both by the centre-right Liberals and the far-right DPP (Vankilde 2013).

The large-scale act was also criticised by members of the Red-Green Alliance as well as by 3F, a major trade union, and the far-right and centre-right parties on the grounds that it constituted "social dumping," the practice of employing migrant labour below the standard of the host country. This view held that the violation of the principal of social dumping placed Denmark in breach of its obligations to the International Labour Organisation (ILO), and as such was a matter of foreign policy over which Denmark had jurisdiction (Vankilde 2012; Hvass 2012).

Without dismissing the possibility that these expressions of concern were genuine, the context of the debates suggests that the criticisms should be considered first and foremost as tactically effective attacks on the governing coalition, which had only been in government for less than a year after replacing the previous centre-right coalition led by the Liberal government of Lars Løkke Rasmussen. In addition, the Greenland issue offered politically useful common ground between opposition parties from opposite ends of the political spectrum.

As a political narrative, the criticisms of the Large-Scale Projects Act effectively combined labour issues, national security concerns, human rights and democracy, nationalist xenophobia and anti-immigration sentiment, all while painting the governing coalition led by the Social Democrat party as ineptly selling out national assets to foreign powers. The issue, and the concerns raised in connection with the issue, quite remarkably threaded a line through policy positions important to each of the major opposition parties. Additionally, the centre-right Liberal Party's opposition to Greenland's mining projects developed only after the party joined the opposition. Prior to the September 2011 parliamentary elections in Denmark, Greenland's Inuit Ataqatigiit (IA) party had been seeking mining investments from China for more than a year, and plans had been in place for many years prior to that to finance autonomy through mining projects. The London Mining Company had been active in Greenland since at least 2008. The Liberal Party government of Anders Fogh Rasmussen had been the primary interlocutor of the Greenland authorities regarding the 2008 self-government referendum, and these plans did not arouse controversy at the time.

What seems to have happened, then, is that the opposition parties in Denmark's parliament mounted an effective campaign against the governing coalition's management of Greenland's transition to self-government. The debates about Greenland were not framed around the basic self-government framework, either because there was no opposition to this or because this line of criticism would have put the Liberals who took part in the talks that produced it in an awkward position.

Instead, a narrative about Chinese interests in Iceland evolved into a story about the threat China posed to labour standards and migration policy, and this in turn fed back into the wider narrative of the security threat posed by China to Danish interests in the Arctic. These debates amplified the perception in Denmark and elsewhere of China as a major player in Arctic, a view that was relatively rare in discussions of Arctic security before 2012 but now features regularly in commentary on the region (e.g. Degeorges 2016; Economy 2014; Pincus and Ali 2015).

Conclusion

The above account of the Greenland controversy argues that Denmark's domestic political debates heavily influenced the Danish response to Chinese foreign direct investments (FDI) in Greenland, and that these debates interacted with a narrative of Chinese interests that was already emerging due to a private Chinese businessman's bid to purchase land in Iceland. This chapter does not argue that China does not have strategic interests in the region, rather it argues that nothing in the controversy over Greenland provides any evidence to support a view of Chinese ambitions in the Arctic region, and there is no evidence to support claims that Chinese investment in the region is part a campaign to win influence or position military assets in the region. The nature of Danish parliamentary politics between 2011 and 2014 provides a much simpler and sufficient explanation of the controversy surrounding mining in Greenland and the wide attention paid to the subject of Chinese investment in Greenland's mining industry.

The Greenland case is particularly striking because of the disconnect between the intense rhetoric surrounding the threat posed by Chinese investment and the empirical reality that no Chinese investment in any major mine in Greenland was on the table during this period. There are projects in the pipeline in Greenland that involve partnerships with Chinese engineering companies, and it appears likely that future expansions of those projects will involve more cooperation with those partners. As discussed earlier in this chapter, Chinese mining companies have been among the most active investors in global mining acquisitions and greenfield investments, and it would therefore not be surprising if Chinese investment in a major mine were to materialise at some point in the future. But for the moment, and in particular during the period of this study, there has not been enough investment to warrant a serious debate.

More generally, the role of domestic discourse in setting and sustaining states' perceptions of security threats is overlooked and is likely important in many other cases in which Chinese investors have been assumed to be acting strategically on behalf of the state. The case in Iceland, detailed only briefly in this chapter, is worth closer attention in this regard. The general lack of transparency regarding the ownership and operations of Chinese firms and the lack of information about individual investors, together with the dominant narrative that presents China as a rising power and a challenger to the existing international order, certainly exacerbates the tendency to view Chinese investments as strategically motivated. The account presented in this chapter, however, suggests that in the case of Greenland there is no evidence to support this narrative.

With respect to Greenland's own political status and economic development strategy, the Danish reaction to Greenland's plans for attracting investment from China presents a nettlesome policy challenge for Greenland. Greenland is of geopolitical importance to Denmark, and while Denmark retains authority over foreign and security policy, the scope of this term has been a point of contention. This presents an opening for political actors in Denmark who may seek to challenge the extent of Greenland's self-government by raising the security implications of various economic development issues; Greenland is particularly vulnerable to such challenges due to its presumed geopolitical significance.

What, in practical terms, does this mean for Greenland's future? Greenland's efforts to attract foreign capital for economic development may again become the subject of political controversy in Denmark. The above discussion suggests, however, that this depends more on coalitional politics in Denmark than on fundamental concerns regarding security policy. This means, for example, that foreign investment in Greenland is less likely to arouse political controversy in Denmark when a conservative coalition is in government, as has been the case since the June 2015 general election.

Acknowledgements

I would like to thank Tim Boersma, Jon Rahbek-Clemmensen, Kristian Søby Kristensen, Steven Ward, and an anonymous reviewer, as well as the co-authors and fellow participants in a conference at the Copenhagen University Center for Military Studies in November 2015 for helpful comments at various stages of this project. All mistakes are my own.

Notes

1 This is not to say that there has been no Chinese activity at all in Greenland. A group of Chinese geologists has been engaged in exploration activities in remote parts of eastern Greenland for several years, and a Chinese engineering company has entered into engineering and distribution contracts with two Australian mining companies on projects that have not yet obtained final licences at the time of writing. Finally, the claim to the Isua project was purchased by a Hong Kong-based company in 2015 after the London Mining Company went into administration. This investment, however, appears to be purely financial.
2 I use the term "elite discourse" loosely here to refer to comments by politicians and reports and commentary from the Danish political press, and to distinguish it from public opinion and the beliefs of the government.
3 All interviews referenced in this chapter were conducted in 2014 as part of research for Boersma and Foley (2014). No new interviews were conducted for this chapter.

References

Adams, W. (2011). A blow to Europe's far-right: Denmark reshapes its immigration policies. *Time*. Available at: http://http://j.mp/2bWVInF. [Accessed 12 March 2017].
BBC. (2011). China's Huang Nubo seeks Iceland land for eco-resort. Available at: http://goo.gl/tQJUnw. [Accessed 11 March 2017].
Boersma, T. and Foley, K. (2014). *The Greenland Gold Rush: Promise and Pitfalls of Greenland's Energy and Mineral Resources*. Washington, DC: Brookings Institution.
Boersma, T. and Foley, K. (2015). Dark clouds gather over Greenland's mining ambitions. Brookings Institution Planet Policy Blog. Available at: http://j.mp/2bMgMwr. [Accessed 11 March 2017].
Danish Defence Intelligence Service. (2011–2016). *Intelligence Risk Assessment* (series). Available at: http://j.mp/1OjcM58. [Accessed 11 March 2017].

Degeorges, D. (2016). China vs. USA: After South China Sea, the Arctic as a Second Act. Institut Français des Relations Internationales. Available at: http://goo.gl/8JhaaZ. [Accessed 12 March 2017].

Economic Council of Greenland. (2013). *Grønlands Økonomi 2013*. Nuuk: Government of Greenland.

Economy, E. (2014). The four drivers of Beijing's emerging Arctic play and what the world needs to know. *Forbes*. Available at: http://goo.gl/YDYi9p. [Accessed 12 March 2017].

Ellemann-Jensen, U. (2011). Dr. No på vulkaner. *Berlingske*. Available at: http://goo.gl/jeAp1T. [Accessed 11 March 2017].

Guðlaugsson, B. (2011). Kínversk risafjárfesting er takmörkunum háð. *Morgunblaðið*. Available at: http://goo.gl/OdgpF1. [Accessed 12 March 2017].

Halskov, L. and Davidsen-Nielsen, H. (2012). Kina vil tjene milliarder på råstoffer i Grønlands undergrund. *Politiken*. Available at: http://goo.gl/374ri8. [Accessed 12 March 2017].

Høegh, P. (2012). Mødtes med kinesisk minister i Tianjin. KNR. Available at: http://j.mp/1MYm0At. [Accessed 19 November 2015].

Hvass, J. (2012). Thorning om social dumping på Grønland: Stop den patroniserende tone. *Politiken*. Available at: http://j.mp/1NgJTtd. [Accessed 12 March 2017].

Humphreys, D. (2014). The mining industry and the supply of critical minerals. In: G. Gunn, ed., *Critical Metals Handbook*. London: Wiley-Blackwell, pp. 20–40.

Index Mundi. (2015). Iron ore monthly price – US dollars per dry metric ton. Available at: http://www.indexmundi.com/commodities/?commodity=iron-ore&months=120. [Accessed 12 March 2017].

Jackson, R. and L. Hook. 2011. Iceland rejects Chinese investor's land bid, Financial Times. Available at: https://www.ft.com/content/26b0f8e2-178a-11e1-b157-00144feabdc0. [Accessed 21 May 2017].

Jerichow, A. (2012). Kast ikke Grønland i Kinas favn. *Politiken*. Available at: http://goo.gl/1VSjds. [Accessed 12 March 2017].

Johnston, A. (2013). How new and assertive is China's new assertiveness? *International Security* 37(4), pp. 7–48.

Jun, P. (2012). After year of talks, Sichuan Miner still no closer to Greenland Deal. *Hot Copper*. Available at: https://web.archive.org/web/20170522204614/https:/hotcopper. com.au/threads/article.2137394/#.WSNOJFKZNP0 [Accessed 19 November 2015].

Jun, P. (2013). China's arctic mining adventure left out in the cold. *Caixin Magazine*. Available at: http://j.mp/1T32yrl. [Accessed 11/19/2015].

London Mining Co. (2013). Social impact assessment (SIA). Available at: http://j.mp/2bWVyg3. [Accessed 11 March 2017].

London Mining Co. (2014). Presentation at Toronto PDAC. Available at: http://j.mp/1jaBqd7. [Accessed 19 November 2015].

Ministry for Foreign Affairs of the People's Republic of China. (2005). Zhang Yesui *Fu Buzhang Huijian Danmai Gelinglan Zizhi Zhengfu Zhuxi Ainuokese*n. Available at: http://j.mp/22jLMY3. [Accessed 12 March 2017].

Ministry of Land and Resources of the People's Republic of China. (2011). 2011 nian zhongguo guoji kuangye dahui xiangmu duijie qianyue yishi juxing. Available at: http://j.mp/1UPO41Z. [Accessed 12 March 2017].

Nielsen, J. (2008). Kæmpe aluminiumsværk med grønlandsk vandkraft. *Information*. Available at: http://goo.gl/eo35xM. [Accessed 12 March 2017].

Pincus, R. and Ali, S. (2015). *Diplomacy on Ice: Energy and the Environment in the Arctic and Antarctic*. New Haven, CT: Yale University Press.

PriceWaterhouseCoopers. (2009). *Mining Deals: 2009 Annual Review*. Available at: http://j.mp/2c6vfsq. [Accessed 11 March 2017].

State Oceanic Administration of the People's Republic of China. (2012). Guotu ziyuan bu buzhang xu shaoshi, guojia haiyang ju juzhang liu cigui baihui gelinglan zizhizhengfu zongli kupike keleisite, bing fangwen gelinglan ziran ziyuan yanjiusuo he gelinglan daxue. Available at: https://web.archive.org/web/20170522205149/http:/www.soa.gov. cn/xw/hyyw_90/201211/t20121112_357.html. [Accessed 11 March 2017].

Vahl, B. and Kleemann, N. (2014). *Greenland in Figures 2014*. Nuuk: Statistics Greenland.

Vankilde, J. (2012). Danmark siger ja til social dumping i Grønland. *Politiken*. Available at: http://j.mp/1lyWrQn. [Accessed 12 March 2017].

Vankilde, J. (2013). Flertal: Staten bør gå ind i Grønland. *Politiken*. Available at: http://j. mp/1PErEhI. [Accessed 12 March 2017].

Wang, N. (2012). Sikkerhedspolitik i Arktis: en ligning med mange ubekendte. *Sikkerhedspolitisk Info*. Available at: https://krigsvidenskab.dk/sikkerhedspolitik-i-arktis-en-ligning-med-mange-ubekendte [Accessed 21 May 2017].

Winther, S. (2013). Løkke vil have timeout: Er dybt bekymret over kinesiske lønninger i Grønland. *Politiken*. Available at: http://j.mp/1I3S7NW. [Accessed 12 March 2017].

Yonhap News Agency. (2012). S. Korea, Greenland agree to cooperate in green growth, resources development. Available at: http://j.mp/1O7zzTf. [Accessed 19 November 2015].

8 The divergent scalar strategies of the Greenlandic government and the Inuit Circumpolar Council

Hannes Gerhardt

When considering the political status and future of Greenland, it is instructive to consider how its paradiplomacy relates to the wider transnational efforts of the circumpolar Inuit community more broadly. It should not be forgotten that Greenland, as a territory, is composed of different ethnicities, Danes and Inuit, of which the Inuit are the clear majority. In this sense Inuit Greenlanders could see themselves as belonging to multiple nations, Inuit, Greenlandic, or Danish. It should be noted that similar scenarios have played out in the establishment of post-colonial states, yet what is particularly interesting in this case are the divergent scalar dynamics involved in nation and state construction. I hope to show that the Inuit Circumpolar Council (ICC) and the government of Greenland engage in quite different scalar constructions, or deconstructions, of nationhood and statehood, with each seeking to further their own particular political agenda.

To begin, the ideas of "state" and "sovereignty" are problematic concepts that need to be unpacked. For one, the belief in the geographical bounding of the state, as a power contained within a particular territory, is fraught with problems and contradictions, as John Agnew (1994) has long argued. The issue of the territorialisation of the state is further problematised by the question of the ontological status of scale more generally. In other words, what is the underlying reality of the power of the state scale over, for instance, regional and local powers? The nature of scale has long been an object of debate in geographical theory and a brief summary will help to elucidate the various strategies employed by Inuit with regard to a Western political ontology centred on state sovereignty.

Original engagements with the concept of scale in geography accepted scales as actually existing levels of spatial organisation, often directly linked to the reach or jurisdiction of economic or political powers. As this conception of scale was also often very hierarchical, real causative power was often seen to lay mostly in the upper scales, the national , and increasingly the global, with subsequent calls made for the need on the part of political operatives to "jump scale", for instance from the local to the national, in order to be effective (Smith 1993). More recently geographers have been increasingly critical towards the concept of "scale", with some arguing it is nothing more than an organising heuristic that does not actually exist in reality; rather the researcher or political activist who proclaims to be engaging scale is actually imposing an artificial template onto the world that obfuscates reality and cripples the effectiveness of political action. In 2005, Marston, Jones and Woodward argued for the expulsion of "scale" as a term, favouring instead a flat ontology that views specific interconnected and emergent sites as a more suitable framework for not only understanding the world, but also changing it.

There have been numerous critical responses to this suggestion that scale should be abandoned altogether. Moore (2008), for instance, proposed that the issue is primarily one of confusing a category of practice with one of analysis. Moore uses the creation of nationhood as a comparable example; a nation, Moore concedes, is a discursive and performative construction and so its reality should not be assumed. Instead, analysis must fall on how and why the nation is constructed the way that it is. According to Moore, this is precisely what must be done with regard to scale, that is, not to assume scalar realities, but to analyse how and for what reasons scalar divisions are created.

In this chapter I will essentially take Moore's lead, turning to the question of how scale is imagined and used within particular political projects of the Inuit. Is it viewed as real or constructed? Are horizontal or hierarchical politics at work? And what is the consequent relation to state sovereignty that emerges from these imaginaries? Yet before turning to these questions, it is necessary to briefly highlight the heavy weight that the idea of sovereignty has in Western political thought and its relation to indigenous peoples. Going back to Thomas Hobbes' political philosophy, state sovereignty has long been conceived of in the West as more than just a given authority over a particular territory, rather it is a space enabled by a political community that allows for the flourishing of civilised society. Importantly, however, there must also, then, exist an outside to this ordered space, that is, the state of nature where life is nasty, brutish and short. This juxtaposition of a normative inside, or *nomos* as Carl Schmitt (2003) would call it, and a chaotic outside pervades Western political theory, and the scalar predominance of the nation-state is directly related to this imaginary.

Armed with a concept of sovereignty this powerful, it is not surprising that European imperial states quickly came to see the New World as the endangering outside of the *nomos*. As Europe then worked to incorporate the New World into its *nomos* the choice for the native inhabitants was clear – be banned or assimilated; or to put it differently, accept the imposed hierarchy of scale, with the state at its centre, or be banished to a realm of disorder and violence. For the Inuit, assimilation efforts have ranged from the coercive to the coaxing as the state worked to foster a particular subjectivity. Over time, the modern Western state increasingly turned to compromise, offering land titles, oil profits, and greater autonomy as modes of assimilation. A problem in this calculus, however, is that significant portions of native peoples, such as the Inuit, may still entail different subjectivities, including their relationship to their land, kin, history, and work, which may embody a way of being that stubbornly remains outside of the political community on which sovereignty is supposed to be grounded.

Thus, the question of the ontological status of scale, particularly with regard to how it undergirds state sovereignty, is not simply an academic one for the Inuit. It is a question that goes to the very heart of their own identity. As I will come to show, however, this identity is not a given. Hence, the Inuit within the ICC and within the government of Greenland position themselves very differently in terms of the scalar imaginaries they embrace and challenge, even as these positions tend to shift and overlap depending on the political challenges being tackled, whether paradiplomatically or from the grassroots.

The ICC and Greenland – different scalar imaginaries

On the face of it, the government of Greenland and the ICC offer fundamentally different strategies to come to terms with the issue of a Western state sovereignty that has largely been superimposed on the lands that the Inuit have inhabited for millennia. While the ICC generally questions the inherent legitimacy of this sovereignty, Greenland seeks to co-opt it by claiming their own state. These different approaches are very much evident in the forms of nationalism that accompany their divergent political projects. While the ICC builds its identity on the existence of a common cultural bond existing between an Inuit people that stretch across the entire circumpolar region, the government of Greenland emphasises a distinct Greenlandic national identity that is not reduced to a particular ethnic group (Nuttall 1994; 2008). In other words, on the one hand we have the construction of a transnational identity, used to achieve greater power over local governance, while on the other we have the construction of a bounded national identity, seeking a territorially based state sovereignty. Clearly there are two very different scalar imaginaries and strategies at work here.

To illustrate these differences we can consider the opposing approaches taken by the ICC and the government of Greenland with regard to sovereignty claims being made on the Arctic Ocean via the UN's Commission on the Limits of the Continental Shelf. The roots of this divergence can be traced back to the Ilulissat deliberations in 2008, where the five Arctic rim states famously declared their pre-eminence in the region based on their boundaries with the Arctic Ocean. There were significant grumblings about this meeting by parties excluded from this declaration, particularly non-rim Arctic states and regional NGOs, such as the ICC (Gillies 2010; Breum 2011). It is worth noting that Greenland was not only a supporter of the Illulisat Declaration, but that they, together with Denmark, were one of the hosts of this meeting, organising, among other things, the invitation list. Greenland subsequently lobbied Denmark hard to make as big a claim as possible on the Arctic Ocean based on the continental shelf that extends from its shores (Breum 2015).

The Greenlandic government is here clearly pursuing a paradiplomacy emphasising a "national" interest in the sovereignty claims issue. This interest is undoubtedly grounded in the potential of Greenland one day becoming an independent state, at which time they would inherit the ocean claims made by Denmark. Thus, while Denmark was weighing several concerns in making their claim, not least their diplomatic relations with Canada and Russia, the Greenlandic government was concerned primarily with achieving the most fortuitous outcome in terms of actual land (or rather ocean) claimed (Breum 2015). There is an irony here in that the narrowness of pursuit with which the Greenlanders approached this question of national interest was made possible by the fact that Greenland is currently not a sovereign nation-state, and hence has far fewer diplomatic considerations to take into account. In the end, Denmark eventually did make an expansive claim that included the North Pole, going all the way to the Russian exclusive economic zone. It should be noted, however, that the role of Greenland's lobbying in this decision is unclear. The strength of the scientific data, colonial guilt/responsibility, and memories of ceding resource rich North Sea territory to Norway in the 1960s may all have played a role in the Danish decision to make the claim that it did (BBC News 2014; Ramskov 2014; Milne 2014).

For the Government of Greenland, however, the paradiplomacy it has employed surrounding the issue of the continental shelf claim has clearly been that of an aspiring sovereign state. There is thus very little emphasis on the transnational Inuit identity that is fundamental to the ICC's political strategy. Rather, Greenlandic politicians are working off of the premise of a Greenlandic nationalism that is rooted in a place-based identity, rather than an ethnic identity (a Greenlander is officially anyone born in Greenland). It must be understood that the foundation of Greenland's paradiplomacy here is grounded in an understanding of self-determination as it harkens back to the Atlantic Charter of 1941 (and later incorporated into the Charter of the United Nations), in which nations are said to have the right to choose their sovereignty. The calculation of achieving full sovereignty on the part of the Greenlandic government can thus be seen to entail a very particular scalar strategy, which, in the end, embraces and reinforces the dominant Westphalian political ontology as it is based on a demand to bounded, sovereign territory within the larger international community of equally bounded and sovereign entities. Inherent in this embrace is the implicit adoption of the standard Western scalar imaginary, that is, a hierarchical ordering of the world which creates distinct levels of authority, with the state still at the apex and entities "below" the state being inherently subservient.

A concern that can be raised on the part of Inuit who are seeking to maintain an alternative way of life, such as the ICC, is that the adoption of the Western political ontology, although intended to create greater freedom from the decisions of a distant capital, may ultimately be a form of assimilation into a non-Inuit worldview (see for instance Boldt and Long 1985). In other words, not only is the discourse and practice of the Greenlandic government actively helping to maintain a scalar "reality" that has long been seen by Inuit across the Arctic as a tool to disempower the native inhabitants, but by claiming the hierarchical standing of the state the claimant begins to act like a state, meaning a circumvention of horizontal, bottom-up political organisation. In the process, the critique goes, the exact same decisions that undermined the Inuit way of life from afar will continue to be made even if they are now coming from a capital that is somewhat closer. A good example of such a critique pertains to the Greenlandic government's decision in 2013 to lift a decades-long prohibition against mining uranium and rare earth minerals that result in uranium by-products (Areddy and Bomsdorf 2013). ICC Greenland, for instance, has long been critical of lifting the ban, not only for environmental reasons, but also because of the impact it could have on traditional ways of life and livelihoods

(ICC Greenland 2013). In its decision, Nuuk is seen as a classic Western capital as it seeks ways to make the Arctic profitable for the sake of narrow national economic priorities, which have come to trump any local or regional concerns.

Turning then to the ICC we can discern a very different political actor. The ICC was formed in 1977, focused on cultivating the ties between Inuit across state borders with the aim of giving a political voice to the Inuit within these states. Consequently, the issue of state sovereignty, and the scalar imaginary that enables it, has been a fundamental focus. On the one hand, and I will return to this shortly, the ICC has fought to ensure that the Inuit are granted their full rights as enfranchised citizens within their respective states, despite being a small minority. On the other hand, the ICC has also fundamentally embraced their identity as an indigenous people with certain entitlements that supersede their state-given rights. It is this emphasis on the Inuit as a distinctive non-Western people, with claims that exist outside of the *nomos* established within a given sovereign space, that places the ICC at potential loggerheads with the Western state and its imperative to assimilate (Corntassel and Primeau 1995).

It is instructive here to consider the ICC's Declaration on Sovereignty from 2009, which was a direct response to the Ilulissat meeting. In this document the ICC highlights the constructed nature of state sovereignty, stating, "sovereignty is a contested concept … and does not have a fixed meaning" (2009). Thus, by not accepting state sovereignty as a taken for granted concept the ICC manages to partially shift the terms of their political relationship with the imposing state. Instead the ICC has squarely focused on a much more nuanced understanding of self-determination than the one embraced by the government of Greenland. Rather than emphasising sovereignty, this alternative view of self-determination, which draws from various UN texts including the International Labour Organization's Indigenous and Tribal People's Convention (1989) and the Declaration on the Rights of Indigenous Peoples (2007), entails a view of an unbreakable tie to specific home territories that cannot be invalidated by any external claims. Critically, such a focus entails a well-honed emphasis on Inuit identity, including their specific forms of livelihood, spirituality, and knowledge, all of which are seen to exist fundamentally outside of the state's imperative of assimilation.

Related to this dismissive stance on sovereignty, the ICC has openly questioned the legitimacy of the continental shelf claims made by the Arctic rim states. According to some key Inuit actors, large swathes of ocean that are being claimed by these states are considered to be part of the Inuit homeland, or *Inuit Nunaat*. *Referencing* the UN's Declaration on the Rights of Indigenous Peoples which asserts indigenous peoples rights to their lands *and* waters, the ICC and affiliated groups have frankly stated that any claims on these waters that ignore the already existing rights to this space by the Inuit have no legitimacy (ICC 2014). In its 2014 General Assembly declaration the ICC committed to work towards helping "Inuit to promote and protect Inuit … rights to the offshore" (ICC 2014). This refusal to acquiesce to the scalar imaginary that legitimises the sovereignty claims of a distant capital over the waters of *Inuit Nunaat* could here be seen as the embrace of a "flat" ontology in which the scalar organisation of hierarchical levels of political authority is seen as a spurious construction that hides and overpowers the complex reality of the networked localised sites that are actually shaping the place of the Arctic. As we will see, however, this assessment is only part of the full picture.

The need for adjusting the adopted scalar strategy

As established above, the scalar imaginaries and strategies of the ICC and the Greenlandic government appear to be quite distinct, yet in what follows I intend to show how both of these political actors also diverge from their themes, tweaking them in order to achieve the best results. To begin, let us keep our focus on the ICC. Despite the political discourse described above, it would be incorrect to conclude that the ICC's scalar imaginary is purely horizontal – while they may argue that hierarchical scales that bestow sovereignty are constructed and heuristically deficient, they nonetheless are also very aware that this imaginary is more than a frame with which to understand the world; it is an instrument of control that must be acknowledged and dealt with. That is why at the same time that the ICC questions the legitimacy of the state-based sovereignty imposed from afar, they also seek to insert themselves in this sovereignty claim as a way to ensure that their interests are not completely ignored.

To achieve this the ICC has chosen to highlight a fundamental and internationally recognised criteria for sovereignty's validity, namely that the land being claimed is inhabited by people that belong to that nation-state. Returning to the ICC Declaration on Sovereignty in the Arctic, the document states, "The foundation, projection and enjoyment of Arctic sovereignty and sovereign rights all require healthy and sustainable communities in the Arctic. In this sense, "sovereignty begins at home" (ICC 2009). In other words, in order to claim that a piece of the Arctic falls under Canada's sovereign rule, for instance, the inhabitants of this land must be Canadian citizens. The ICC can then go on to assert, if the Inuit are to consider themselves Canadian citizens, Inuit rights must be recognised as well. In other words, the self-determination that is laid out in the UN's Declaration on the Rights of Indigenous Peoples is used not so much as a call to outright resistance, but rather as a tool to gain a greater voice in the governance process.

In this way the Inuit of the ICC are reaching out and offering a compromise – acknowledging the scalar imaginary used to make sovereignty claims in exchange for a seat at the table. It is precisely such a strategy that has led to land claims agreements in the US and Canada, and to increased autonomy for Greenland. More recently the ICC has taken a similar approach with regard to sovereignty declarations over the Arctic Ocean. Thus, after calling to defend Inuit "sovereignty" in the 2014 ICC General Assembly declaration, the ICC goes on to urge Inuit involvement in any continental shelf claim submission, "to involve Inuit in those submissions so as to reflect the Inuit perspective and protect Inuit rights and interests". In this sense the ICC recognises that the claims being made on Inuit land, and now water, are backed by considerable power, and hence the political strategy is to ensure that this power is tamed through the direct involvement of the Inuit.

What we can conclude from the above is that the ICC's emphasis on the constructed nature of hierarchical scale does not actually lead to a purely flat ontology in which state sovereignty is denied in favour of bottom-up, horizontal political organisation. When push comes to shove in determining who decides in the Arctic, as with impending sovereignty claims on the Arctic Ocean, the ICC's imperative has been to ensure that they have a voice. Working to establish this voice has led the ICC to seek forums that stretch beyond the confines of the national discussion, particularly by working with transnational organisations such as the United Nations and the Arctic Council (Shadian 2010). In this way the ICC may be recognising the staying power of a state-centred, Western political ontology, but they are doing so in a way that continues to insist that any claims to sovereignty in the Arctic depend on recognition from the Inuit on the local scale as well as from the international community on the global scale. Indeed, the state, while being recognised by the ICC, is not being recognised as a reified thing; rather it is seen as a continual process and practice, within which, at least in the Arctic, the Inuit must be viewed as an integral part (Steinberg et al. 2015).

Turning now to the government of Greenland, we can see that they also have been pushed to reposition their scalar strategy, at least on the surface, when it comes to the all-in stake that they seem to have taken with regard to state sovereignty. To understand such a shift it is necessary to consider the challenges Greenland has had in maintaining a scalar imaginary that sees itself as both inside and outside of the Danish sovereign realm. This challenge became very clear in Greenland's spat with Denmark and the Arctic Council in 2013.

In the past Greenland and the Faroe Islands were accustomed to being a distinct and separate part of the Danish delegation in official Arctic Council deliberations. This arrangement ended after the Arctic Council's leadership passed to Sweden in 2011. In response, the newly elected premier of Greenland at that time, Aleqa Hammond, decided to boycott the Arctic Council meeting in Kiruna in protest. Yet the move on the part of the Greenlandic premier revealed a contradiction at the heart of their implicit embrace of a Western political ontology.

Ultimately, there appear to be two options for justifying a demand for special Greenlandic representation at the Arctic Council, and, for that matter any other forums that assemble state leaders. One is to claim, much as the ICC does, that the indigenous inhabitants of an area have the right to be represented in decision-making that pertains to these inhabitants. Here, the sovereign rule of the state is made contingent on incorporating these voices. This is undoubtedly how Danish officials have viewed the rights granted to Greenland and the Faroe Islands. If this were the argument on the part of the Greenlandic government, however, then the claim could not be made that Greenland should be given special treatment over indigenous Arctic peoples' NGOs. For instance, why should Greenland be represented at the Ilulissat deliberations, or deliberations over continental shelf claims, but not Inuit representatives from the ICC?

The other option for Greenland is to be considered a sovereign entity within a sovereign entity, essentially claiming to be inside yet outside of Denmark's sovereign domain. This appears to be the approach that Greenland has often taken. As such, Greenland is essentially asking to be considered a state, or more accurately to be considered a quasi-state, or a state-in-the-making. Such a claim, however, flies in the face of an embrace of the Western scalar imaginary, in which you either have sovereignty or you do not. In other words one gets the impression that Greenland embraces the scalar imaginary that justifies its territorial and absolute claim to sovereignty upon achieving independence, yet at the same time denies Denmark the right to this claim in the interim.

While Greenland appears to be comfortable with this position, other states and indigenous groups are not. Greenland's boycott of the Arctic Council was immediately cast by other states as a Danish internal affair. To them Greenland still clearly falls within Danish jurisdiction, the scalar hierarchy is thus maintained. The ICC, in turn, also appeared miffed with the Greenlandic government's strategy. Although no official statements seem to have been made, Aqluuk Lynge, a Greenlander and member of the ICC's old guard, made no bones about criticising the premier's move, questioning its effectiveness and viewing it as undermining an institution (Arctic Council) that the ICC holds dear (George 2014). For the ICC, it should be noted, the compromise of recognising state sovereignty in exchange for political rights is considered an integral component of the Arctic Council. It is clear that the actions of the Greenlandic government were thus not viewed by the ICC as a call to arms on the part of the transnational Inuit cause, rather it was a political move on the part of an aspiring nation-state.

There is a final irony to this story. In the end the Greenlandic government was able to win its seat back – mostly due to efforts on the part of the Danish government, which was eager to diffuse the embarrassing development (Breum 2015). Yet in the process the Greenlanders also lost their right to display their flag at Arctic Council deliberations. The premier, who weathered significant criticism for her decision from many fronts, spun the result as a victory for the Greenlandic people (Nunatsiaq News 2013). And perhaps it was. Yet one is left to wonder whether this was the result that the premier was truly after – if the protest was one of seeking greater influence on the deliberations by having a seat next to the Danes, then it was a victory, but if the paradiplomatic protest was about asserting ones nationalistic right to be at the table, based on a recognition of quasi-statehood, then the loss of the flag, the ultimate symbol of that state, appears to be a loss.

Scalar overlap?

Perhaps it is the frustration of not being recognised as a quasi-sovereign state that has led the Greenlandic government to make a strategic shift in its scalar strategy in the recent United Nations Conference on Climate in Paris, where the Greenlandic government, despite being part of the Danish delegation, decided also to partner with the ICC and Nunavut in making a joint declaration. In this declaration a call was made to limit global warming to a maximum of two degrees Celsius, emphasising that warming most greatly impacts the way the peoples of the Arctic "live and work". Many of the standard demands made by the ICC are also reflected, particularly the recognition and protection of "the rights of

Indigenous peoples and the values, interests, culture and traditions of the peoples of the Arctic" (Government of Greenland et al. 2015). The basis for these demands are also clearly influenced by the ICC, namely that the lands most effected by global warming are the homeland of the Inuit, and any policymaking that affects this region requires the input of the local inhabitants. For Greenland to sign such a document, however, is interesting, as the Greenlandic government has long been promoting a vision of future development, based on energy and mining, that generally is recognised as benefitting from global warming while leading to non-traditional forms of "living and working" (Nuttall 2008).

Yet to understand Greenland's participation in such a document, we must look at its other components. The document also distinguishes between the responsibilities of the developed and underdeveloped world with regard to global climate change; it calls for the right to development; and it requests assistance from the industrialised world to adapt to the challenges of climate change. All of these are ultimately economic, not cultural demands that the Greenlandic government was hoping to bring attention to with the help of the ICC and Nunavut. In fact, Vittus Qujaukitsoq, Minister of Finance, Mineral Resources, and Foreign Affairs, who headed the Greenlandic team at the Paris summit, stated that his primary aim was to ensure the right to continued growth in Greenland, particularly within the energy and mining sector. Mr. Qujaukitsoq explicitly mentioned plans for ten new mines in Greenland that he was eager to protect from any commitment that Denmark might make as part of the climate agreement (Holten-Møller 2015).

The Greenlandic government's scalar strategy here is quite nuanced. On the one hand Greenland's paradiplomatic demands echo those of a developing state, seeking the right to develop its economy without "unfair" restrictions imposed by the wealthier industrialised nations. Here the sovereignty of Denmark is again questioned as Greenland presents itself as its own quasi-sovereign entity. Indeed, the Greenlandic government ultimately made clear that they would not consider themselves a part of the deal, despite Denmark's signature, as it would prevent them from developing their nascent fossil-fuel industry. Yet on the other hand, Greenland also presented its case couched in the language most familiar to the ICC, that is, with a focus on the rights of the local indigenous peoples. As the official statement from Mr. Qujaukitsoq read:

> Our joint Inuit voice and our traditional know-how from across the Arctic should be heard and included in international policymaking. Most importantly, Arctic indigenous peoples have to be ensured an equal access to the right to development.
>
> (Government of Greenland et al. 2015)

This is certainly a departure from past discourses of Greenlandic nationalism that have shied away from embracing Inuit ethnic identity. The scalar dynamics are noteworthy: Greenland, faced with the possible imposition of a sovereign authority beyond their territory turn to use transnational institutions (the UN and the ICC) to demand local/regional rights, all the while looking out for their one primary goal, maintaining the ability to develop in order to one day have the resources needed to claim full sovereignty via the establishment of an independent state.

Conclusion

What we learn from the above examples is that there is a fluid relation on the part of the ICC and the Greenlandic government to the current Western construct of scale, and the institution of state sovereignty that it enables. While both of these political actors ultimately stake out different scalar strategies in the pursuit of their political aims, these strategies are tweaked and nuanced to fit the challenges and opportunities of a given moment. For its part, the ICC tends to focus on the still counter-hegemonic, non-assimilated aspects of the Inuit's life-world. Thus the ICC generally plays up its identity as a rallying call against the outright dominance of Western sovereignty, which it portrays as a questionable construct. At the same time, the ICC recognises the reality of the power that is attached to this construct, and in order to have at least some say in the decisions affecting their homeland, the ICC chooses to acknowledge the imposed scalar hierarchy on the condition that the Inuit are made part of Arctic governance.

Greenland, in turn, generally tends to de-emphasise its Inuit identity as it tries to play the political game on the Western sovereign's terms. The goal here is clear, to play the game in order to win by achieving full state sovereignty. While the hierarchical scalar order of the Western political ontology has thus been embraced and legitimised, the Greenlandic government has sometimes found itself in the paradoxical position of demanding to be recognised as a quasi-sovereign entity, both inside and outside of Danish sovereign jurisdiction. Finding it difficult to always achieve traction with such an approach, Greenland has been shown to also make use of the claim to indigenous rights, much as the ICC, which at its root actually questions the taken for granted legitimacy of scalar sovereignty.

Yet despite the occasional overlap of scalar strategies between the Greenlandic government's paradiplomacy and the ICC's political activism, the underlying rift between the two runs more deeply. It is a rift that exists between two separate visions, two distinct approaches that the Inuit people can take with regard to Western sovereignty and the nomos to which it is attached. The choice is essentially one of calculated assimilation into the hegemonic Western world or pragmatically executed resistance to it. It is very possible that both of these approaches will come increasingly into conflict with each other within the Greenlandic political realm. If statehood is achieved, Greenland will likely become part of a larger logic, behaving and making decisions that will be hard to distinguish from those made from afar prior to independence. For Greenland, then, this could mean the blossoming of a much more vibrant civil society, in which the ICC and other organisations come to the fore. Thus, instead of an Inuit state that adopts the call of the indigenous rights of the native Arctic people, Greenland may become just another product of the Western scalar hierarchy, a state within which political entities, such as the ICC, demand a greater say in the decisions that affect the people on the ground, dispersed among Greenlands' many remote localities.

References

Agnew, J. (1994). The territorial trap: the geographical assumptions of international relations theory. *Review of International Political Economy* 1(1), pp. 53–80.

Areddy, J., and Bomsdorf, C. (2013). Greenland Opens Door to Mining. *Wall Street Journal*. Available at: www.wsj.com/articles/no-headline-available-1382712417 [Accessed 21 May 2017].

BBC. (2014). Denmark challenges Russia and Canada over North Pole. Available at: http://www.bbc.com/news/world-europe-30481309. [Accessed 12 March 2017].

Boldt, M., and Long, J. (1985). Tribal philosophies and the Canadian Charter of Rights and Freedoms. In: M. Boldt and J. Anthony, eds, *The Quest for Justice: Aboriginal Peoples and Aboriginal Rights*. Toronto: University of Toronto Press, pp. 165–179.

Breum, M. (2011). *Når isen forsvinder: Danmark som stormagt i Arktis, olien i Grønland og kampen om Nordpolen*. Copenhagen: Gyldendal.

Breum, M. (2015). *The Greenland Dilemma. The Quest for Independence, the Underground Riches and the Troubled Relations with Denmark*. Copenhagen: Royal Danish Defense College.

Corntassel, J. and Primeau, T. (1995). Indigenous sovereignty and international law: Revised strategies for pursuing 'self-determination'. *Human Rights Quarterly* 17(2), pp. 343–365.

George, J. (2014). Arctic Council indigenous orgs rally for secretariat's 20th anniversary. *Nunatsiaq Online*. Available at: http://www.nunatsiaqonline.ca/stories/article/65674arctic_councils_permanent_participants_rally_for_their_secretariats_20/. [Accessed 12 March 2017].

Gillies, R. (2010). Clinton decries exclusions from Arctic meeting. *Associated Press*. Available at: http://www.sfgate.com/news/article/Clinton-decries-exclusions-from-Arctic-meeting-3193532.php [Accessed 21 May 2017].

Government of Greenland, Government of Nunavut, and ICC. (2015). Governments of Nunavut and Greenland, and Inuit Circumpolar Council issue joint statement on climate change. Press release. Available at: http://www.gov.nu.ca/eia/news/governments-nunavut-and-greenland-and-inuit-circumpolar-council-issue-joint-statement. [Accessed 12 March 2017].

Holten-Møller, C. (2015). Grønland går efter vækst og minedrift på klimatopmøde. *Kalaallit Nunaata Radioa*. Available at: http://knr.gl/da/nyheder/gr%C3%B8nland-g%C3%A5r-efter-v%C3%A6kst-og-minedrift-p%C3%A5-klimatopm%C3%B8de. [Accessed 12 March 2017].

ICC. (2009). A circumpolar Inuit declaration on sovereignty in the Arctic. Available at: http://www.inuitcircumpolar.com/sovereignty-in-the-arctic.html. [Accessed 12 March 2017].

ICC. (2014). Kitigaaryuit Declaration. Available at: http://www.inuitcircumpolar.com/uploads/3/0/5/4/30542564/img-724172331.pdf. [Accessed 12 March 2017].

ICC Greenland. (2013). Appeal to the Greenlandic and Danish governments not to abolish the uranium zero-tolerance policy in the Danish realm. Available at: https://www.nirs.org/wp-content/uploads/international/westerne/Statement%20on%20uranium%20mining%20in%20Greenland%2026%20April.pdf. [Accessed 12 March 2017].

Marston, S., Jones J., and Woodward K. (2005). Human geography without scale. *Transactions of the Institute of British Geographers* 30(4), pp. 416–32.

Moore, A. (2008). Rethinking scale as a geographical category: From analysis to practice. *Progress in Human Geography* 32(2), pp. 203–225.

Milne, R. (2014). Denmark lays formal claim to North Pole. *Financial Times*. Available at: https://www.ft.com/content/b7e66b1c-8442-11e4-8cc5-00144feabdc0 [Accessed 21 May 2017].

Nunatsiaq News. (2013). Greenland ends boycott, returns to the Arctic Council. Available at: http://www.nunatsiaqonline.ca/stories/article/65674greenland_comes_back_to_the_arctic_council/. [Accessed 12 March 2017].

Nuttall, M. (1994). Greenland: Emergence of an Inuit homeland. In: Minority Rights Group, ed., *Polar Peoples: Self-Determination and Development*. London: Minority Rights Group, pp. 1–28.

Nuttall, M. (2008). Climate change and the warming politics of autonomy in Greenland. *Indigenous Affairs* 1(2), pp. 44–51.

Ramskov, J. (2014). Derfor gør Danmark nu krav på Nordpolen. *Ingeniøren*. Available at: https://ing.dk/artikel/derfor-goer-danmark-nu-krav-paa-nordpolen-172976. [Accessed 12 March 2017].

Schmitt, C. (2003). *The Nomos of the Earth in the International Law of the Jus Publicum Europaeum*. New York: Telos Press.

Shadian, J. (2010). From states to polities: Re-conceptualizing sovereignty through Inuit governance. *European Journal of International Relations* 16(3), pp. 485–510.

Smith, N. (1993). Homeless/global: scaling places. In: J. Bird, B. Curtis, T. Putnam, G. Robertson and L. Tickner, eds, *Mapping the Futures*. London: Routledge, pp. 87–119.

Steinberg, P., Tasch, J., and Gerhardt, H. (2015). *Contesting the Arctic: Politics and Imaginaries in the Circumpolar North*. London: I.B. Tauris.

9 Greenland and the Arctic Council

Subnational regions in a time of Arctic Westphalianisation[1]

Inuuteq Holm Olsen and Jessica M. Shadian

It happened over lunch. It was Tuesday 15 May 2012. The senior Arctic officials (SAOs) of the Arctic Council[2] were in Stockholm for a deputy minister's meeting, when it was unveiled that Greenland and the Faroe Islands would no longer have a spot at the negotiating table alongside the Danish SAO (Denmark is an Arctic country by virtue of its political relations with Greenland and the Faroe Islands). In reality, the lunch was only one instance among a growing number of changes taking place within the Arctic Council. The origins of these changes reach back to the 2009 Arctic Council Ministerial meeting in Tromsø, Norway. Unlike the low-profile nature of past meetings, the 2009 Arctic Council meeting included political leaders from former United States vice president, Al Gore to the Chinese minister to Norway and Michel Rocard (former Prime Minister of France under François Mitterrand and appointed French ambassador to the Arctic and Antarctic), among others.

Underlying the impetus behind the sudden increased global attention to the Arctic and subsequent attendance at the 2009 meeting was due to the fact that the ministerial meeting was the first official meeting following the planting of a tiny titanium Russian flag on the seabed at the North Pole by Artur Chilingarov (Russia's most famous Arctic explorer) in a submarine accompanied by a fellow parliamentarian, a Swedish businessman and an Australian tour operator. Since that flag planting, global interest in the Arctic has grown exponentially. Powerful global states including China, Japan, South Korea, the EU, India and Singapore have set out – with great success for most parties – to become permanent observers on the Arctic Council (the EU has been less successful). In attempts to circumvent, rather than overtly dismantling, the privileged position of the six indigenous permanent participants (PPs) who also sit at the negotiating table alongside the Arctic states, the Arctic Council responded to growing interest through varying actions that resemble conventional intergovernmental politics (politics by and for states). This includes a growing number of binding agreements that have been negotiated by the eight Arctic states (with various levels – from some to none – of participation by the permanent participant organisations) under the auspices of the Arctic Council (the Arctic Council does not have the authority to make legally binding agreements).

Parallel to those changes taking place in and around the Arctic Council, Arctic politics more broadly has been affected by emerging global political trends. Beginning in the 1970s, a number of subnational and transnational Arctic governments and institutions have been increasingly forging ahead with their own Arctic politics and collaborations, at times operating on a global scale. In certain instances, their efforts bypass national governments and often those collaborations fall outside of the scope of the Arctic Council altogether.

This chapter will focus on the future of the Arctic Council in light of this renewed global interest in the Arctic alongside the rise of globally situated subnational Arctic regions. In particular, this chapter will focus on a global Greenland as a window into the incongruent forces between the Westphalianisation of the Arctic Council and the increasing institutionalisation and assertion of subnational Arctic politics. This chapter will begin by laying out a brief narrative of what we refer to as the Westphalianisation of the Arctic Council, before moving on to a theoretical discussion about regions and the politics underlying the borders that create them. The theoretical backdrop provides context for the following section which looks specifically at a changing Arctic through the eyes of Greenland before taking a closer look at the intersection between the forces of sub-national regions and a Westphalianising Arctic Council.

Westphalianisation of the Arctic Council

The founding of the United Nations in October 1945 reaffirmed the concept of state sovereignty through the creation of modern international law. Writing at the time of the UN's founding, almost 300 years after the Peace of Westphalia, Leo Gross noted that while the political map had changed greatly since the treaty, its chief political idea had "undergone relatively little change" (Gross 1948, 21). In modern international law, sovereignty entailed a highly specific conception of "nation" – one that was closely related to territory (i.e. to territorial integrity). Because of the long-standing premise that land was something that could be owned and exchanged, nationhood also became a legal aspiration, which was to attain territorial integrity (Rudolph 2005, 127). In effect, territorial integrity became a precondition of international standing. According to Anghie (1999, 62),

> sovereignty represents at the most basic level an assertion of power and authority, a means by which a people may preserve and assert their distinctive culture. …
> For the non-European society, personhood as recognised internationally was achieved precisely when the society ceased to have an independent existence, when it was absorbed into European colonial empires or when it profoundly altered its own cultural practices and political organisations.

Fast-forwarding to the end of the Cold War and the era which followed, global politics had – by the end of the century – manifested into a global politics marked by the creation and resurgence of sub- and non-state polities that very often physically and politically transcended the long entrenched idea of the Westphalian political system. The new political terrain became that which was not defined by a waning of nationalism and nationalist movements or the demise of the nation-state system, but rather by what Rosenau (1997, 243) referred to as "fragmegration" or "resistances to boundary-spanning activities" which also act simultaneously with the rise of new orders and institutions or integration.

In the midst of these changes the first iterations of the Arctic Council was conceived and eventually brought to fruition. When the Ottawa Agreement was finally signed in 1996, the Arctic Council established itself as a high-level forum to: "provide a means for promoting cooperation, coordination and interaction among the Arctic States, with the involvement of Arctic indigenous communities and other Arctic inhabitants on common Arctic issues" (Arctic Council 1996). The new political collaboration established a platform for the eight Arctic states and three indigenous permanent participant organisations (which are now six) to sit together at the negotiating table to discuss how the Arctic environment should be preserved, developed, and governed for the benefit of the Arctic states and those living there. Well beyond the exclusion of security matters, the Arctic Council was namely established as a means to discuss Arctic environmental protection and sustainable development. Given the impossibilities to create a political regime during the Cold War, establishing full circumpolar cooperation of any nature in the early post-Cold War years was deemed as a great accomplishment.

One significant moment leading up to the creation of the Arctic Council was Mikhail Gorbachev's 1987 address to the international community from the Arctic city of Murmansk. At that meeting he called on the Arctic to become an international "zone of peace" (Gorbachev 1987, 9). That meeting has now become symbolic for initiating a fundamental shift in the history of Arctic politics. Whereas up until that point the Arctic was long recognised by the international community as either a no man's land (the final frontier); strategically, as the place where the East met the West during the Cold War; or considered significant for scientific purposes, Gorbachev's speech initiated a new process of political region-building. The speech laid down the process which led to the creation of the Arctic Environmental Protection Strategy (AEPS) that is the predecessor to the Arctic Council.

Two decades later, another prominent Russian remade the role and significance of the Arctic in global politics. Chilingarov's Russian flag planting on the seabed at the North Pole once again instantaneously and irreversibly altered the Arctic's geopolitical relevance. Despite the decades of established region-building, going back to when Gorbachev first foresaw the Arctic as "a zone of peace", much of the world viewed Chilingarov's flag planting as an act of Russian aggression in a region without governance and still frozen in a Cold War politics (Doward et al. 2007; Live Leak 2009; Borgerson 2008). Shortly following the flag planting, the eight Arctic states, six PPs, and a surge of non-Arctic politicians gathered in the Arctic Norwegian city of Tromsø to attend the 2009 ministerial meeting of the Arctic Council. The newly attending non-Arctic states wanted to learn more about the Arctic Council and moreover how they can become involved in this "new" global geopolitics of the Arctic. The Arctic states, taken by surprise at the numbers of newly interested non-Arctic states and organisations (for example, the International Association of Oil and Gas Producers submitted an application to become an observer) found themselves suddenly faced with a new reality that the Arctic Council had become a global political regime and that the time had come to revisit its mandate and its goals. Void of the ability to make international law, the main question for the Arctic states was to determine what role the Arctic Council should play in the governance of the region, who should be part of this governance, and moreover where the Arctic Council sits in the realm of global politics going into the future.

Since 2009, the direction that the Arctic Council has adopted is to start making its way down the path of conventional Westphalian politics, coming just short of fully realising the necessities that come along with that move. Due to new interest and "pressure" by non-Arctic countries to be more involved in Arctic governance, the Arctic Council member states have begun to re-align the way the council has traditionally operated. For instance, a number of Arctic countries began meeting outside of the Arctic Council and therefore without the permanent participants. Likewise, some Arctic states began to impose controls which would only recognise and allow the original member states to sit around the table and make decisions. The consequences of the latter action eliminated the practise whereby, until 2011, Greenland and the Faroe Islands sat at the table alongside Denmark. Despite not being full members of the Arctic Council, Greenland and the Faroe Islands are the principal Arctic actors in the Kingdom of Denmark's Arctic affairs.

Further, since its founding, the Arctic Council has transitioned from a "high-level forum" (Arctic Council 1996) at its inception into a "high level *intergovernmental* forum" (Arctic Council n.d.a. emphasis added) to provide a means for promoting cooperation, coordination and interaction among the Arctic states, with the involvement of the Arctic Indigenous communities and other Arctic inhabitants (Arctic Council n.d.a.). In international law *intergovernmental* specifically refers to an institution (forum or otherwise) comprised of sovereign states. It then becomes an intergovernmental *organisation* through the establishment of a Treaty. Without the existence of a treaty, however, the Arctic Council cannot make formal (hard law) international policies (Harvard Law School n.d.).

Despite these shortcomings and recognising its inability to take legal action, since 2011 the Arctic states – "under the auspices of" (and not as a directive of) the Arctic Council – have produced a number of binding agreements. This is something historically unique for the Arctic Council member states. The binding agreements began with the 2011 "Agreement on Cooperation in Aeronautical and Maritime Search and Rescue in the Arctic" (SAR) which was followed by the "Agreement on Cooperation on Marine Oil Pollution, Preparedness and Response in the Arctic" (Oil-Spill Agreement) in 2013. Then in October 2015 the Arctic Coast Guard Forum was established and several other binding agreements are currently under negotiation (including a science cooperation agreement).

The move towards binding agreements is significant in a number of ways. The first is the obvious point that the agreements are indications of the fact that – despite that the Arctic Council does not have legal authority to make international law – the Arctic states have decided that they do want to be able to determine how Arctic governance should proceed (rather than leave policy to the UN or another international body). Second, because the agreements were made among the Arctic States and under the auspices of the Arctic Council rather than by the Arctic Council itself, they have largely excluded the six indigenous permanent participants (though the PPs were part of the discussions leading up to the Oil-Spill Agreement). This includes the fact that the PPs did not sign the agreements. These two factors combined are, as such, bringing to the surface the broader question facing the future structure and mandate of the Arctic Council: Who does the Arctic Council serve and who gets to govern?

Though the Arctic Council includes six indigenous organisations (representing Northern indigenous peoples) the new binding agreements are made by the eight Arctic states and those states' capitals are situated in the southern regions of those states, capitals which are often physically located very far from the north and therefore have very different realities and priorities than their northern governments and peoples. Even though it is the northern regions that give the eight states their status as *Arctic* states in the first place, it is often the foreign ministers and civil servants working in capitals to the south who serve as the SAOs and Arctic Council ministers (Canada has a long-standing exemption in this regard which includes member of parliament from Nunavut, Leona Aglukkaq, who served as Canada's minister for the Arctic Council and its chair from 2013 to 2015 as well as Mary Simon and Jack Anawak who have both served as the Canadian ambassador to the Arctic).

Subnational northern regions, therefore, are increasingly finding that they need to go through southern capitals to be heard within the Arctic Council. In effect, the Arctic Council is increasingly speaking on behalf of and is making decisions about its northern regions without their representation. In return, those regions then have little choice but to implement what has been decided. As the Arctic Council continues to Westphalianise and proceed down the path of "intergovernmental" forum, the lingering legacies of colonial practice are coming to the surface. Alaska State House representative from Bethel, AL Bob Herron poignantly remarked on the prevailing colonial mentalities when he had to "fly south" to participate in an Arctic Encounter Symposium in Seattle in January 2016:

> We're not someone's convenient snow globe so they can look inside the snow globe and see all these little fur-clothed, subsistence people living in a zoo, in a museum, in an environment where they must protect it … There's a couple times where I've felt that I've been patted on the head and they've said, "Don't worry. We'll take care of you."
>
> (Miller 2016)

The questions that house representative Herron's comment raises are whether or not, as the Arctic Council evolves, southern powers are once again becoming gatekeepers of the north. Who controls the political narrative of the Arctic? In a region of regions, where do the political and governing borders of the Arctic begin and end?

The Arctic: a region of regions

Unlike Africa or Asia, the northern regions of North America and its indigenous peoples had to wait until the 1970s to begin their own political processes of self-determination. Likewise, what was similar was that all aims for self-determination began with expectations to develop resources on Arctic indigenous lands and the debates which followed eventually led to land claims agreements in Alaska and Canada, various forms of cultural autonomy and indigenous rights in the European Arctic, as well as the Government of Greenland.

Greenlandic home rule came into being in 1979. The impetus behind greater Greenlandic autonomy from Denmark stemmed from the reactions by Greenlanders against Danish policies regarding Greenland during the 1950s to the 1970s.

With the passage of the Home Rule Act, Greenland – unlike the land claims processes in Canada and Alaska – did not give up Inuit title. Likewise, whereas Alaska land claims are focused most strongly on indigenous corporations and the Canadian land claims have been processes of decentralisation of power away from Ottawa to the local land claims settlement regions, Greenland since home rule has slowly gone through a process of transferring control from Denmark to Greenland. Home rule, among an entire host of other inclusions, established a public government. Though it was an issue during the Home Rule Commission, Greenland was not granted the right to control mineral resources. The compromise was the establishment of a joint Danish–Greenland Council on Mineral Resources consisting of five members of parliament from Denmark and Greenland. By doing so, Greenland was given veto power, preventing the Danish government from enacting any essential decisions regarding Greenland without the consent of the Home Rule Government of Greenland or vice versa (Christiansen 2015, 72–73). Likewise, with Greenlandic secession from the EU, Greenland took control of its fisheries but at the same time it entered into a fisheries agreement with the EU. In that agreement, the EU pays for fishing rights in Greenlandic waters.

Since 1979, Greenland has gone through two commissions which have concluded with greater autonomy (the first was a pure Greenlandic commission and the second consisted of equal number of representatives from Denmark and Greenland). Finally, in June 2009, self-government came into effect, giving Greenlanders total control over all surface and subsurface rights from 2010 onwards. As the legislation determined, some resource revenues will go towards paying against the Danish block grant allocated to Greenland every year. While remaining under the Danish realm, Greenland under self-government now owns outright all of its surface and subsurface resources and negotiates internationally on its own accord in areas that both fall under Greenlandic jurisdiction and geographically deal with Greenland itself. Areas such as foreign affairs covering the whole of the kingdom as well as security and defence, however, remain under the Danish authorities. Despite those rules, whenever Greenlandic interest or issues of relevance are involved, Denmark has to include Greenland's interest in those deliberations or negotiations (Gad 2012; Ackrén and Jakobsen 2015, 404–412; Skydsbjerg 1999).

One example of these divided competences is the issue of uranium mining. Uranium is a well-regulated mineral that is covered by several UN conventions and aspects dealing with security and defence. After lengthy negotiations, Greenland and Denmark entered into four agreements on how to handle and proceed with the question of uranium. Having taken over the area of mineral resources, Greenland gives permission for exploration and exploitation licenses for minerals containing uranium while Denmark will have the responsibility of securing the compliance of several UN (especially IAEA) conventions and agreements including in areas of safeguard measures, export control, and non-proliferation. Uranium mining has become the first instance (since the 2009 Self-Rule Act) which has created a clear divide between the competences of the government of Greenland and the responsibility of the international subject, herein the state of Denmark.

Overall, Greenland exemplifies a vastly changing Arctic political landscape. Moreover, as the connections between the Arctic and the global economy strengthen through possibilities of new shipping routes and increased ship traffic across the Arctic, coupled by renewed interest in Arctic resources and a growing tourism industry, subnational Arctic regions are becoming increasingly globalised alongside the domestic changes at home. Unlike earlier periods of history, when explorers and entrepreneurs from around the world came to the Arctic and were met with little if any political resistance, in today's Arctic, subnational entities from Greenland to Nunavut and Alaska have set up institutions of governance, hybrid cooperation, and corporations which collaborate directly with global industry and government as well as monitor and regulate activities.

Greenland alone has a direct and ongoing relationship with the EU since its withdrawal in 1985. Over the years it has evolved such that Greenland now has a permanent representation in Brussels. In September 2014, Greenland opened a North American office in Washington, DC. Both of those offices meet with policymakers, industry, and other associations to establish direct ties and relations with the North American governments, subnational governments, industry, and other entities on behalf of Greenland. Greenland's Brussels and North American representations also serve to disseminate outwards Greenlandic views and interests in the Arctic region and globally.

A changing Arctic Council through the eyes of Greenland

Another major aspect of Greenland's foreign diplomacy is its long standing participation in Arctic cooperation which dates back to the Arctic Environmental Protection Strategy (AEPS) – the precursor to the Arctic Council – including the discussions leading up to its establishment on 14 June 1991. Greenland, subsequently, was a participant throughout the negotiations to establish the Arctic Council. As an Arctic nation, the government of Greenland believes that it is imperative for Greenland to take part in and to contribute to regional policy discussions in a political forum like the Arctic Council, specifically when those decisions affect Greenland and its people.

Leading up to the formation of the Arctic Council, it was considered only logical that Greenlandic policymakers were part of the discussions and negotiations as well as the evolution of its work since its inception. Denmark has also historically recognised the critical role of Greenland during the negotiations and continues to recognise its role more generally on the Arctic Council through today. As an exemplar of this, at the inauguration of the Arctic Council, the then premier of Greenland, Lars Emil Johansen, signed the Ottawa Declaration on behalf of the Kingdom of Denmark. Likewise, in the early years ministers from Greenland were often the head of delegation for Denmark (e.g. the Barrow Declaration in 2000 and Reykjavik Declaration in 2004 (Arctic Council website, n.d.b.)). Further, Greenland has been consistently active in many of the working groups including its role as the lead delegation as well as chair of various working groups. For example, Greenland represents the Kingdom of Denmark in the Sustainable Development Working Group as well as the Protection of Arctic Marine Environment (PAME).

Throughout the 2000s, the Danish delegation to the Arctic Council consisted of the Faroe Islands, Greenland, and Denmark. All political entities participated on equal terms. There were three chairs at the table and all three parties participated in the executive meetings as well as ordinary meetings of the senior Arctic officials. The country label was "Denmark/Faroe Islands/Greenland" and all three flags were prominently displayed at the table. These displays did not consist of a change of the membership status from the Ottawa Declaration, but there was tacit agreement that this was how the Kingdom of Denmark represented itself.

For Denmark, it has been a long-standing practice to include Greenland and the Faroe Islands in all delegations where all three bodies have vested interests. Denmark's practice of conducting foreign policy was not always well understood by other countries' diplomats; its politics differed greatly from the other Arctic countries' own political structures at home and thus their conduct for diplomacy. Yet, when it came to the Arctic Council, the tripartite Danish delegation had become accepted practice.

During the 2011–2013 Swedish chairmanship of the Arctic Council, however, the Westphalianisation of the Arctic began to manifest itself in the operations of the Arctic Council. This change from the status quo began when Greenland and the Faroe Islands suddenly found themselves excluded from the executive SAO meetings – the place where the most high–level political negotiations and decisions are made. The form of exclusion, interestingly enough, came in the form of *chairs* (not a formal letter or other official protocol). Suddenly, the designated spot for the Kingdom of Denmark at the negotiating table went from having three chairs to one chair. Greenland and the Faroe Islands were left to find chairs of their own away from the table (which sometimes included finding chairs located outside of the negotiation room altogether).

That was in 2011 and the exclusion of Greenland and the Faroe Islands continued for the following two years throughout the Swedish chairmanship. The new seating arrangements did not, however, go unchallenged. The dissatisfaction came to a head when Greenland decided to boycott the ministerial meeting in

Kiruna on 15 May 2013. At that time, Greenland further announced that it was suspending all of its ongoing activities with the Arctic Council until a resolution was found. Though a resolution was finally found it was not until the new Canadian chairmanship in 2013–2015 (Government of Greenland 2013c, 8).

The Arctic Council Ministerial meeting in Nuuk, Greenland that concluded the Kingdom of Denmark's chairmanship adopted the Nuuk Declaration of 12 May 2011. The ministers decided on a process to strengthen the Arctic Council by establishing a permanent secretariat as well as creating a task force to look into the rules of procedures under the Swedish chairmanship (Arctic Council 2011a, 2; 2011b, 3). It can only be speculated as to why the Swedish chairmanship decided to do that, but some member states might have seen the opportunity to tighten up the procedures as a means to reduce the role played by Greenland in the work of the Arctic Council.

The period leading up to Greenland's re-engagement with the Arctic Council was driven by a combination of four main factors: the international media attention that Greenland's boycott caused, internal Arctic Council reactions to the boycott, political deliberations by Denmark with the Arctic Council on behalf of Greenland, as well as the extensive debates at home in Greenland about its decision to boycott the Arctic Council and the subsequent ramifications that those actions caused. The chair of the Greenland Parliament's Permanent Committee on Foreign Policy and Security, Per Berthelsen, publicly argued that he had serious doubts that the Canadian chairmanship would be more open to the demands of Greenland. According to Berthelsen,

> Inuit in Canada are a minority. If Greenland achieves direct participation in Arctic Council negotiations, Canada will suddenly be faced with a dilemma. Our [Canadian Inuit] kinsmen will probably demand the same role as Greenland if we are brought in from the cold.
>
> (Mølgaard 2013a)

The opposition leader at that time, former premier Kuupik Kleist criticised the absence of Greenland from the ministerial meeting by noting that:

> [t]he super powers have a whole different agenda. They avert giving indigenous peoples influence by keeping the power themselves. USA's access to the Arctic Council is because of Alaska's position and the Northern provinces in Canada have also given Canada its access to the Arctic Council. They will not let go of their seats at the table in the Arctic Council.
>
> (Mølgaard 2013c)

With the start of the Canadian chairmanship, Greenland together with the Faroe Islands and Denmark set out to negotiate with Canada to find a satisfactory solution to the issue of representation at the executive and ordinary SAO meetings (Government of Greenland 2013a). The negotiations lasted several months and finally on 19 August 2013, the government of Greenland published a press release which announced that an agreement with the Canadian chairmanship had been concluded and that Greenland could resume its participation at the Arctic Council. The concluding arrangement was such that, going into the future, all three political bodies of the Danish Delegation would have full participation rights at the Arctic Council meetings. When the number of seats to each delegation is less than three, the person or persons who would sit at the table would be determined according to which representative in the Kingdom has competence on the matter being discussed.

That decision also falls in line with the Self-Government Act of 2009, which states that Greenland can enter into and negotiate international agreements in matters where it has taken over competence from Denmark on issues that pertain to Greenland and further that Greenland will gradually take over new areas of responsibility (Government of Greenland 2013b). In practical terms, and in the context of the Arctic Council, there is once again tacit consent that when the number of seats for the Kingdom of Denmark is less than three, the delegation will rotate its seat at the negotiating table depending on the subject matter and which delegate has the greatest competence and legitimacy to take part in the discussions.

Despite the new arrangements, not everyone was content. Though Greenland and the Faroe Islands were allowed, once again, to participate at the table of the Artic Council with transition of the chairmanship from Sweden to Canada, it did not come without a new form of exclusion. With the change of chairmanship the small flags that were conventionally placed at the table designated to each participant were taken away (thereby taking away the three flags in the Danish Kingdom) and replaced with large full-sized flags of only the member states and the PPs behind each chair.

The main opposition party in Greenland questioned whether or not the new situation restored the Greenlandic position to its former capacity. In a similar critique, the leader of the main opposition party, Kuupik Kleist, remarked in the Greenlandic press that, at the end of the day, the Kingdom of Denmark only had one vote on the Arctic Council. Kleist went on to point out that:

> [Greenlanders] had preferred to see that the subject matter of the self-governing countries' [of Greenland and the Faroe Islands] role in the Arctic Council be discussed as a separate agenda item during an Arctic Council meeting instead of Greenland going at it alone. The issue is not only about Greenland but encompasses many other Arctic areas.
>
> (Mølgaard 2013b)

This then brings us to the present situation within the Arctic Council. Greenland has resumed its participation and work in the Arctic Council. It has a seat at the table at the SAO meetings as well as in the working groups (due to the internal recognition and flexibility within the delegation of the Kingdom of Denmark). The other Arctic states have attempted to dictate what the delegation of the Kingdom of Denmark should look like (despite that the situation is a domestic Danish issue which falls outside the mandate of the Arctic Council) yet Greenland has acquired the legal capacity at home to make decisions that directly affect Greenlanders. Greenland, as such, has the right to be involved in the work and decision-making processes of the work of the Arctic Council. Nonetheless, the reality is that, for Greenland, the Arctic Council looks increasingly like an intergovernmental regime while at the same time it is increasingly only one venue among a number of emerging platforms for Greenland to engage in Arctic and global politics.

Arctic Council: not the only player in town

Beyond the Arctic Council, increasingly so, new forms of Arctic cooperation are emerging or renewing themselves, which is leaving some commentators to question whether or not such entities are a complement or a possible competition to the Arctic Council (Conley and Melino 2016). A number of examples include Arctic Frontiers, the Arctic Circle Forum, the Arctic Economic Council (AEC), the Northern Forum and the World Economic Forum: Global Agenda Council on the Arctic. In the context of Greenland, successful self-determination is increasingly dependent on engaging with non-Arctic Council entities. The government of Greenland is currently experiencing substantial expectations and responsibilities at the domestic level which are directly tied to its engagements in the regional and global political fora. Self-government has forced Greenland to take responsibility of its own economic development, finances, and essentially economic prosperity. Subsequently, Greenland spends an increasing amount of its energy entering into agreements with global entities, including states and the private sector, as it takes over the former competencies from Denmark. These realities are vastly transforming the political, social, and economic institutions in Greenland.

Conclusion

The Arctic has never quite fitted into the mould of conventional Westphalian political system. When the Arctic Council was created, it was the first regional political organisation to include non-state actors and in many ways, therefore, it served as a harbinger for a world to come. With increasing global interest in the Arctic, the Arctic Council seems to be making efforts to go back in time and become a conventional intergovernmental political regime. Yet, these efforts also come at a time when some argue that the days of conventional formal international law and policy making on its own are numbered. Increasingly so, norms in global governance includes the participation of non-state actors, best practices, and other forms of soft law (Shaffer and Pollack 2010).

If the Arctic Council continues to diminish involvement of the permanent participants, then some of the most critical issues and changes coming to the Arctic are going to fall outside of its mandate. With the growing power of sub-national Arctic regions and new non-Arctic Council entities engaging in Arctic activities, will the Arctic Council be able to keep pace with an increasingly global Arctic (Shadian 2016; Dodds 2016; Heininen 2016)? Is there a space for subnational regions on the Arctic Council? If not, what are the implications of subnational governments doing global negotiations and policymaking concerning the Arctic (trade deals or otherwise) completely separate of the Arctic Council?

Global economic and other interests in the Arctic will continue to grow going into the future. Greenland's own political future is dependent on its economic performance and development, namely assuming complete responsibility for its government (the ability to cover the costs of those sectors such as justice, law enforcement, immigration policy etc.). As a result, global entities apart from the Arctic Council will play an even greater role in Greenlandic politics.

Much like Greenland, the Arctic's sub-national governments will continue to work towards having a greater say in matters that are of interest and relevance to them. Though the Arctic Council states are turning to conventional state politics to strengthen Arctic governance, one might question if those very actions might be undermining its power. Global politics is undergoing vast changes and the Arctic Council needs to decide what role it wants to play in the Arctic going into the future.

Notes

1 For a longer version of this chapter, please see: Shadian, J. and Olsen, I. (2016). Greenland and the Arctic Council: Subnational regions in a time of Arctic Westphalianisation. Arctic Yearbook 5, pp. 229–50. Available at: http://arcticyearbook. com/images/Articles_2016/scholarly-articles/8-AY2016-Shadian.pdf

2 The Arctic Council is the only fully circumpolar political regime for the Arctic led by the eight Arctic states (Russia, Canada, United States, Iceland, Denmark (Greenland and Faroe Islands), Sweden, Norway, and Finland) and six indigenous organisations (Inuit Circumpolar Council (ICC), the Sami Council, The Russian Association of Indigenous Peoples of the North (RAIPON), the Aleut International Association (AIA), The Arctic Athabaskan Council (AAC), and the Gwich'in Council International (GCI). It is considered by many as the official body for conducting circumpolar Arctic governance (e.g. Shadian 2014).

References

Ackrén, M. and Jakobsen, U. (2015). Greenland as a self-governing sub-national territory in international relations: Past, current and future perspectives. *Polar Record* 51(4), pp. 404–412.

Anghie, A. (1999). Finding the peripheries: Sovereignty and colonialism in nineteenth-century international law. *Harvard International Law Journal* 40(1), pp. 1–80.

Arctic Council. (1996). Joint communique of the governments of the Arctic countries on the establishment of the Arctic Council. Available at: https://oaarchive.arctic-council.org/ bitstream/handle/11374/85/EDOCS-1752-v2-ACMMCA00_Ottawa_1996_Founding_ Declaration.PDF?sequence=5&isAllowed=y. [Accessed 11 March 2017].

Arctic Council. (2011a). Nuuk Declaration. Available at: https://oaarchive.arctic-council.org/bitstream/handle/11374/92/07_nuuk_declaration_2011_signed. pdf?sequence=1&isAllowed=y. [Accessed 13 March 2017].

Arctic Council. (2011b). Senior Arctic Officials (SAO) Report to Ministers. Available at: https://oaarchive.arctic-council.org/bitstream/handle/11374/1535/SAO_Report_to_ Ministers_-_Nuuk_Ministerial_Meeting_May_2011.pdf?sequence=1&isAllowed=y. [Accessed 13 March 2017].

Arctic Council. (n.d.a.). Arctic Council Website. Available at: http://www.arctic-council. org/index.php/en/about-us [Last accessed 19 May 2016].

Arctic Council. (n.d.b.). All Arctic Council Declarations 1996-2015. Available at: https:// oaarchive.arctic-council.org/bitstream/handle/11374/94/EDOCS-1200-v3-All_Arctic_ Council_Declarations_1996-2015_Searchable.PDF?sequence=4&isAllowed=y. [Accessed 7 November 2016].

Borgerson, S. (2008). Arctic Meltdown. *Foreign Affairs* 87(2), pp. 63–77.

Christiansen, S. (2015). *Kajs Grønlandskrønike*. Copenhagen: Informations forlag.

Conley, H. and Melino, M. (2016). An Arctic redesign: Recommendations to rejuvenate the Arctic Council. Center for Strategic and International Studies. Available at: https:// www.csis.org/analysis/arctic-redesign. [Accessed 12 October 2016].

Dodds, K. (2016). What do we mean when we talk about the global Arctic? *Arctic Deeply*. Available at: https://www.newsdeeply.com/arctic/community/2016/02/18/what-we-mean-when-we-talk-about-the-global-arctic. [Accessed 11 March 2017].

Doward, J., McKie, R. and Parfitt, T. (2007). Russia leads race for North Pole oil. *The Guardian*. Available at: http://www.theguardian.com/world/2007/jul/29/russia.oil. [Accessed 16 September 2013].

Gad, U. (2012). Greenland projecting sovereignty: Denmark protecting sovereignty away. In: R. Adler-Nissen and U. Gad, ed., *European Integration and Postcolonial Sovereignty Games: The EU Overseas Countries and Territories*. Abingdon: Routledge, pp. 217–34.

Gorbachev, M. (1987). Speech delivered at the ceremonial meeting on the occasion of the presentation of the Order of Lenin and the Gold Star Medal to the city of Murmansk. Council on Foreign Relations. Available at: http://www.cfr.org/arctic/general-secretary-gorbachevs-speech-murmansk-october-1987/p32441. [Accessed 11 March 2017].

Government of Greenland. (2013a). Danmark, Grønland og Færøernes deltagelse i Arktisk Råd samarbejdet. Press Release. Available at: http://naalakkersuisut.gl/da/ Naalakkersuisut/Nyheder/2013/05/ArktiskRaad. [Accessed 11 March 2017].

Government of Greenland. (2013b). Grønland genoptager sin deltagelse i Arktisk Råd. Press Release. Available at: http://naalakkersuisut.gl/da/Naalakkersuisut/Nyheder/2013/08/ Arktisk-Raad. [Accessed 11 March 2017].

Government of Greenland. (2013c). Udenrigspolitisk Redegørelse 2013. Available at: http:// naalakkersuisut.gl/~/media/Nanoq/Files/Attached%20Files/Udenrigsdirektoratet/DK/ Udenrigspolitiske%20redegorelser/Udenrigspolitisk%20Redeg%C3%B8relse%20 2013.pdf. [Accessed 21 May 2017].

Gross, L. (1948). The Peace of Westphalia, 1648–1948. *The American Journal of International Law* 42(1), pp. 20–41.

Harvard Law School. (n.d.). Intergovernmental organizations (IGOs), Available at: http://hls.harvard.edu/dept/opia/what-is-public-interest-law/public-international-law/ intergovernmental-organizations-igos/. [Accessed 11 March 2017].

Heininen, L. (Ed.). (2016). *Future Security of the Global Arctic: State Policy, Economic Security and Climate*. London: Palgrave Macmillan.

Live Leak. (2009). Military tensions heating up on Canada's coldest frontier. Available at: www.liveleak.com/view?i=8c7_1235958275 [Accessed 17 September 2013].

Miller, M. (2016). Alaskans fly south for Arctic Symposium. KTOO. Available at: http://www. alaskapublic.org/2016/01/15/alaskans-fly-south-for-arctic-symposium/ [11 March 2017].

Mølgaard, N. (2013a). Grønland er bombet flere år tilbage. *Sermitsiaq*. Available at: http:// sermitsiaq.ag/groenland-bombet-flere-aar-tilbage. [Accessed 11 March 2017].

Mølgaard, N. (2013b). IA: Dårlig aftale med Arktisk Råd. *Sermitsiaq*. Available at: http:// sermitsiaq.ag/iadaarlig-aftale-arktisk-raad. [Accessed 11 March 2017].

Mølgaard, N. (2013c). Kuupik: Uklogt af Aleqa. *Sermitsiaq*. Available at: http://sermitsiaq.ag/kuupik-uklogt-aleqa. [Accessed 11 March 2017].

Rosenau, J. (1997). *Along the Domestic-Foreign Frontier: Exploring Governance in a Turbulent World*. Cambridge: Cambridge University Press.

Rudolph, C. (2005). Sovereignty and territorial borders in a global age. *International Studies Review* 7(1), pp. 1–20.

Shaffer G. and Pollack, M. (2010). Hard vs. soft law: Alternatives, complements, and antagonists in international governance. *Minnesota Law Review* 94 (3), pp. 706–799.

Shadian, J. (2016). Finding the global Arctic. *The Arctic Journal*. Available at: http://arcticjournal.com/opinion/2216/finding-global-arctic. [Accessed 11 March 2017].

Skydsbjerg, H. (1999). *Grønland: 20 år med hjemmestyre*. Nuuk: Atuagkat.

10 Materialising Greenland within a critical Arctic geopolitics

Klaus Dodds and Mark Nuttall

> Greenland is an enormous hunk of ice, three times as big as Texas, with a narrow fringe along the southern shore where a few Eskimos and fewer Caucasians scratch out an existence. During World War II the United States spread a protecting wing over this inhospitable territory … Greenland groans under a ponderous icecap that leaves only a slim margin of land sticking out around the edges.
>
> (Roucek 1951, 239)

In 1951, the American sociologist Joseph Roucek (1951) penned an essay entitled "The geopolitics of Greenland" for the *Journal of Geography*. Although not formally trained in geography, Roucek was, for the next forty years, an enthusiastic producer of short articles purporting to chart and track the geopolitics of the Arctic and the Antarctic, as well as other places such as central and eastern Europe and the Mediterranean. Indeed, in another piece, "The geopolitics of the Arctic", published in 1983, he drew attention to the region's potentially rich resources and its strategic military significance as an air route and waterway, referring to "Arctic fuels" as a way of reducing North American dependence on oil from the Arabian Gulf. "The 'Arctic Mediterranean'", he wrote, "is a perfect example of an area in which technological advances, especially in aviation, have caused far-reaching changes which force a new evaluation of locational factors of the region" (Roucek 1983, 463).

In "The geopolitics of Greenland", Roucek was at pains to make information and knowledge about the geographical location, size and geomorphological features of the island accessible to a North American readership, but in doing so he ignored the human–environmental relations nurtured and enacted in the dynamic surroundings of an ever-changing Arctic, over several thousand years. Roucek simplified its people as those only intent on "scratching out an existence" on the margins of an enormous ice sheet. There is little sense of Greenland as both a lively and lived space, in which human life engages with the more than human entities which constitute that world, and where local, national and global connections, dialogues and forces coalesce and collide. For Roucek, Greenland is, to echo Anna Tsing's (2005) phrasing, a place without "friction".

The notion of an icecap being "ponderous" chimes serendipitously with recent work on how glaciers are part of local and indigenous worldviews and are subject to cultural framing (Cruikshank 2004; Orlove et.al. 2008), although Roucek may not have had this in mind. To him, the Greenland ice sheet appears to be a burden – the land "groans" under its weight and the mass of ice emphasises the inhospitable nature of this Arctic territory. And yet that very weight and extent of ice also signals strategic possibility for an extra-territorial party, namely the United States. Roucek was one of many commentators of the time in whose work one finds an evocation of the "desert-like" nature of Greenland, a space with meagre resources making it appear empty and devoid of potential for sustained habitation to non-residents (pace Said 1993). Geopolitically, though, Roucek's attitude towards Greenland reflects a prevailing view that later American interest in Greenland was shaped by the experiences of World War II, and the move by the Danish administration in 1941 (or more appropriately, the Danish ambassador to Washington Henrik Kauffmann) to sign a defence agreement and allow an American presence in Greenland. In his Greenland essay, Roucek also pointed to the strategic importance of air routes and waterways, which he reiterated in "The geopolitics of the Arctic". Greenland's strategic importance during World War II lay not in the possibilities it provided for a northwards mapping and discovery, or access to its resources, but in its position on a North Atlantic stepping stone route for bombers and, critically, for its role in weather forecasting. Several American installations were built during the war, including three air bases – at Narsarsuaq in south Greenland, Kangerlussuaq on the west coast, and Ikateq near Ammassalik on the east coast.

As Roucek observed, Greenland's ice sheet was also attracting renewed scientific interest, especially in its role in northern hemisphere climate patterns. "Scientists," he wrote, "have long suspected that Greenland's icecap manufactures much of the bad weather that sweeps over Europe and perhaps the entire northern hemisphere. But to verify this theory, they needed on-the-spot reports of icecap weather conditions". An object of scientific enquiry since the nineteenth century, Janet Martin-Nielsen describes how the ice sheet became central to Cold War science diplomacy (Martin-Nielsen 2013). Interest in the ice sheet's age, thickness and history was closely related to wartime and Cold War strategic concerns with weather, ocean currents and sea ice. This blending of the scientific and strategic and the geopolitical and geophysical was to be further nourished by new expeditions, notably the Expéditions Polaires Françaises (EPF), and later US military investment in glaciological research in Greenland (Martin-Nielsen 2012).

Our chapter is a material, volumetric and discursive intervention into, onto and across Greenland including its ice mass and surrounding seas. It is not a friction-free encounter, but one where the "geo" in geopolitics is scrutinised. Our advocacy of a critical Arctic geopolitics is one rooted in materiality where the Arctic is not simply a backdrop to human events. Rather we advance an interest in how the materiality of the waters, ice, snow, rock, wind and air of the Arctic becomes available for further geopolitical manifestations. As Elizabeth Grosz (2008) has written on the subject of geopower, the Arctic might be conceived as something that also challenges and even subverts the geopolitical, cartographic and scientific.

Following on from that discussion of a critical Arctic geopolitics, we explore how Greenland's ice sheet was an essential, if at times recalcitrant, accomplice to US and Danish Cold War geopolitical performances and practices. An environment in other words that was capable of challenging and undermining the materials, sites and modalities actors such as the US military brought to bear on it. We then move on to consider another form of materiality and what might be a volumetric geopolitics, in this case the Arctic Ocean seabed and the efforts by the Geological Survey of Denmark and Greenland (GEUS) to map and chart outer continental shelves, including those stretching towards the North Pole (Dodds 2010; Strandsbjerg 2012). Seabed mapping off Greenland's continental shelf illustrates how this accompanies claims for sovereign rights to be extended hundreds of miles from Greenland's coastline. In both cases, separated by five decades, was the issue of how far down (i.e. how thick) and how far out (i.e. how wide) did Greenland actually extend. Greenland as territory, as a consequence, exhibits and expresses itself, as a process rather than outcome and an unstable volume rather than a static and flat surface. To reinforce this point about Greenland being in a continuous state of becoming (Nuttall 2015), we finish with a consideration of the creation of resource spaces and the Greenland frontier. This explicitly material and volumetric accounting of Greenland and its geopolitics show how we might approach a more critical form of Arctic geopolitics, emphasising the vitality of the "geo" in the discursive qualities of Arctic geopolitics (Clark 2013).

Going volumetric: critical Arctic geopolitics

Before we drill into Greenland's icy core or descend into its depth-like qualities, we contemplate this chapter's analytical optic. Rising interest in the Arctic's (changing) geographical qualities informs much of which is to follow. From speculation over the future of Arctic sea ice and its complete disappearance (at least in the summer season) to reflection over the region's resource potential, many scholars and commentators have mused on various Arctic futures. It is now taken for granted that the region is geographically dynamic and it has been framed, mapped, imagined and projected in a myriad of ways, many of which resonate with current concern with humans as agents of geophysical transformation and rupture. The Arctic as resource frontier, endangered homeland, unique ecosystem under threat, epicentre of and for climate change and zone of great power rivalries and rising international interest are just some of the framings used in this conversation about regional futures. As Brunn and Medby conclude:

> Petroleum potentials, mineral riches, shipping lanes, and national strategies are often at the fore of geopolitical accounts of the circumpolar North, but Arctic spatiality can by no means be reduced to the sum of these parts. "The Arctic" is many different things at once: a frontier, a homeland, a highway, a stage, a laboratory. It is a space that has intrigued people for centuries and continues to do so today.
>
> (Bruun and Medby 2014, 915)

We argue that the Arctic might also be thought of in explicitly volumetric terms and, by peering within, above and around and by taking notice of subsurface and ocean depths, mountain and glacial interiors, as well as the atmosphere, thus build on recent scholarship by geographers that challenge "horizontalism" within social science research, neglecting the vertical and depth-like qualities of social and political life (for example, Bridge 2013; Elden 2013). Rachael Squire, in her analysis of Gibraltar, shows how the disputed United Kingdom overseas territory has been locked into an elemental struggle with Spain that encompasses more than the surface (Squire 2016). The seabed and offshore marine environment have been enrolled in rival sovereignty and security projects. In advocating a volumetric approach, this tranche of work reminds us that the Arctic also has the capacity to be filled, to expand and to contract dependent on earthly and human forces, claims, ambitions, ideas and interventions. An environment where digging through rock, chipping away at ice, drilling into glacial depths, navigating within, through, and under polar waters, flying through and across, as well as observing and monitoring Arctic skies. This has profound implications for the scale, scope and intensity of human interventions from pursuing whales and seals, excavating coal and minerals, traversing across, and through ice and establishing routes for aviation.

In our book *The Scramble for the Poles*, we take up this challenge (Dodds and Nuttall 2016). For us, a critical Arctic geopolitics is defined as one attentive to the discursive and representational qualities of its subject matter, but also adoptive of a relational understanding of the world, which in turn is attentive to the connections between human and non-human elements. We therefore advocate a view of the Arctic as a lively space characterised by agency, change, and vitality. Our use of the word "scramble" was intended as provocation to highlight historical associations and representations of the Arctic with earlier "scrambles" for knowledge, appropriation of territory, colonisation of peoples, administration, resources and transportation. The framing of the Arctic as "resource frontier" or "super maritime highway" provides a particular historical trope for journalists and popular writers, as well as academics and policymakers. And it also marks attempts by human agents to stabilise, to exploit, to move through and to appropriate the Arctic as a place composed of ice, sea, air, rock, animals, architectures, landforms and people – with varying degrees of success.

The Arctic has attracted, and continues to attract, the language and imaginative framings of colonial expansion and settlement (and Greenland has certainly not been exempt from such language and framings), and to this we add that Arctic spaces are also lucrative and material sites for speculative capitalism. The resource potential of the Arctic has been actual and imagined. Animal furs and pelts and whale oil proved commercially lucrative in the earlier stages of that colonial European encounter, while ambitions to extract minerals and oil and gas dominate contemporary narratives concerning the "opening up of the Arctic". While, however, there are numerous megaprojects around the circumpolar North concerned with the extraction and processing of minerals and hydrocarbons, there are many more at the planning stages, especially in Greenland, and although some are likely to be implemented, it may well be that for some corporations the Arctic is more important as a frontier for speculative ventures rather than a space for actual resource extraction.

In this way, an extractive industry is successful in how it can raise "promissory capital", as Charis Thompson (2005) terms it, i.e. capital raised on the promise of future returns and, in the case of mining and oil development, interventions in the subterranean and underground. Capital is thus not accumulated – and indeed, projects do not need to undergo construction and operational phases and resources do not necessarily have to be extracted, but the idea of extraction and the hype surrounding it becomes part of a political economy concerned with the reproduction of speculation. Projects, such as the various mining ventures at the exploratory stages in Greenland for instance, become important, assume a political life and a social presence, become central to how politicians and business leaders imagine the future, and are made into capital by virtue of what future success and profits they promise. Imagining the Arctic as a resource frontier may bring the future into the present though its narrative of promise and economic development, but it also brings apprehension and anxiety, especially to those indigenous communities who do not feel they have sufficient information or have not been consulted adequately about a project (Nuttall 2010). So a critical Arctic geopolitics would be attentive to the affective dimensions of social–material relationships and networks enveloping places like Greenland as promising, hopeful and rewarding. But in order to be so, we contend that a substantial body of knowledge produced on Greenland was emblematic of particular forms of Arctic geopolitics, emphasising both the depths and widths of Greenland.

Extracting the subterranean: the Greenland ice sheet and Cold War geopolitics

The US has long expressed a strategic interest in Greenland and other northern regions. As secretary of state in Abraham Lincoln's administration, William Henry Seward argued the US needed to have both Greenland and Alaska within its national borders so that it could exercise sovereignty over the North Pacific and North Atlantic, and thus control the approaches to the North American Arctic (Hough 2013). Alaska was purchased from Russia in 1867 and Seward continued to eye Greenland as well as Iceland. A military rationale partly inspired this desire, but if the territories could become American possessions Seward saw opportunities for exploration, mapping and ownership of resources in areas already claimed by Russia and British Canada. Seward could not garner sufficient interest in Washington DC to make a formal approach to the Danes, but his ideas supported exploration in the High Arctic. Following the tragedy of Adolphus Greely's Arctic expedition of 1881–84 (Greely survived, but most of his crew perished), however, the US government ended its financing of Arctic exploration, and American expeditions were largely funded by private sponsors and geographical exploration societies for the next forty years or so (Robinson 2006). It is also worth noting that, in 1916, during negotiations for the transfer (or more accurately, the sale) of the Danish West Indies to the United States, the

Americans accepted a demand from Copenhagen that they would not object to the extension of Danish sovereignty over the whole of Greenland.

In the aftermath of World War II, US strategic planners recognised that Greenland and the wider Arctic region mattered to hemispheric security. As they lie on the shortest route for a possible Soviet attack on the North American mainland, islands such as Greenland were crucial for the construction of military surveillance and transport systems in the Arctic. Detailed knowledge of the Arctic's landscapes and seascapes was essential as planners needed better understanding of the impact and effect of permafrost, sea ice, glacial ice and prevailing weather systems on road construction and maintenance, air flights, ship and submarine mobility, navigation and tracking. The Greenland ice sheet was one of many elements of the Arctic under scientific and strategic scrutiny. Martin-Nielsen has discussed, for example, how between 1948 and 1966 US forces in Greenland were entrenched in the "other cold war". This was a struggle with the environment; the ice was a formidable opponent in how it acted to impede American ambitions in the High Arctic. Martin-Nielsen argues that the Americans faced two choices: they could either approach the Greenland environment, and its ice sheet in particular, as something to conquer and control, or they could choose to enter into a relationship based on strategic cooperation. It was the latter approach which was chosen (Martin-Nielsen 2012).

Control of the North Atlantic was crucial for military and strategic advantage during the war and this depended on having accurate meteorological knowledge. Knowing what the weather was like in Greenland and the northern North Atlantic was vital for knowing what the weather would be like in northwest Europe a few days later. Greenland was placed at the centre of an assemblage of military and scientific technology and infrastructure, as airbases were constructed and manned by several thousand military personnel, and as weather stations measured atmospheric conditions. This Arctic was not, then, seen in horizontal terms as a space in which to enter, explore, traverse, and map. Greenland became important to how we look up into the sky and the atmosphere. Air routes and the scientific-technological mapping and measuring of northern spaces and the control of meteorological knowledge placed Greenland in a new global system in which a volumetric geopolitics was enacted.

Post-1945, the US–Danish relationship was strengthened by the decision by Denmark to join NATO. At the time when Roucek was writing, in 1951, the US military presence in Greenland was being consolidated by the Danish–American defence agreement in the face of concerns about Soviet intentions, and the importance of Greenland for aerial routes and maritime surveillance in the North Atlantic and Arctic Ocean (Dunbar 1950; Petersen 1998). Thule Air Base was established at Pituffik in 1951 and entailed the forced relocation of the Inughuit living there to Qaanaaq, 140km north, in 1953. Whereas during World War II Greenland's strategic role lay in linking North America and Europe, and in weather forecasting, in the Cold War this role was redefined into a strategic aerial base for the US against the Soviet Union, and later Denmark secretly acquiesced to let the US station nuclear weapons at Thule.

As the geographer Isaiah Bowman noted in 1949, "Survey, survey, and survey may be said to be the three basic requirements of present-day polar research, and we do not restrict the word to cartography" (Bowman 1949). As the Expéditions Polaires Françaises demonstrated, new endeavours were brought to bear on Greenland's ice using tracked vehicles, airplanes and new scientific instruments. Between 1949–1951, the EPF carried out hundreds of seismic and gravitational readings, carried out altitude measurements, and used theodolites to survey and map the extent of the ice sheet. The EPF's work also represents some of the earliest examples of ice coring. EPF scientists in collaboration with the Danish Geodaetisk Institute were also interested in the mass balance of the ice sheet, the relationship between accumulation and ablation and thus overall ice sheet stability. The end result was a different kind of mapping of Greenland, emphasising the volumetric. An emphasis addressing the depth, the surface and volume of the ice sheet. In addition to the dissemination of the purely scientific data, maps of the profile and interior depths of Greenland were published, most spectacularly in 1956 via the *National Geographic* magazine, for public audiences.

Using cutting-edge scientific techniques, supported by an extensive logistical programme, and well-versed in public engagement, the research raised the profile of Greenland within the popular global imagination. However, growing strategic interest in the Arctic was something the Soviet Union and United States shared in the early Cold War. Soviet interest in sea ice to the north of the Russian Arctic coastline was matched by American investment in the ice-filled environment of the North American Arctic, including Greenland, and so framing the island as a "bastion" for the defence of the North American continent. As historians of Cold War science and technology have noted, investment in snow and ice research followed and the US Army's Snow, Ice and Permafrost Research Establishment (SIPRE), which had been founded in 1949, was moved to Wilmette in Illinois in 1951. In 1953, Project Mint Julep was launched to investigate whether the southern area of the Greenland ice sheet could be used to support aircraft landing strips. Further north, US engineers were constructing the airbase at Thule to develop and sustain distance early warning capabilities, in the event of a sneak attack by Soviet bombers.

Further, SIPRE in their Operation Icecap, probed beneath the surface of the ice sheet in the north west of Greenland. SIPRE scientists, working closely with US Army personnel, were leading research on the movement of glacial ice, the properties of snow and ice, and the stability of glacial masses such as the Greenlandic and Antarctic ice sheets. Remarkably, Greenlandic ice was being transported to cold rooms at the SIPRE headquarters. Once there, scientists began the task of archiving the ice, measuring accumulation layers and probing the physical and chemical composition of the ice core.[1]

The calculative and investigative qualities of this ice core work had important implications for Arctic geopolitics. By the late 1950s, the US had established a series of sites for military and scientific investigation and operation including Thule Air Base, Camp Tuto (18 miles from the main Thule complex), Site 2 (further east in the interior of northwest Greenland) and Camp Century (the so-called "City under the Ice"). Inspired in part by the 1957–8 International Geophysical Year, Camp Century was intended officially to be a test site for subsurface engineering. The engineering rationale for Camp Century was actually a cover-story for a more sinister project called Ice Worm, designed to facilitate an extraordinary complex of tracks and tunnels to store, hide and deploy nuclear missiles.[2] Accordingly, the Greenlandic ice sheet became enrolled and implicated in a strategic investment to use the dynamic qualities of snow and ice to conceal US missiles. The material geographies of Greenland, its ice, seabed and snow, and its rocks and minerals, are central to any geopolitical auditing of Cold War Arctic geopolitics and beyond. The interior of Greenland also became bound up with a popular culture that amazed North American and European audiences, as the US Army revealed in its work when building the "city under the ice" at Camp Century (SIPRE scientists were key participants in the project) (Kinney 2013).

The lively materiality and movement of ice proved to be the downfall of Project Ice Worm. By the mid-1960s, ice deformation was compromising the safety of the tunnel network and the growing importance of US submarines as a mobile missile nuclear force compromised its strategic value. The digging, moving, manipulating, and managing of mobile ice overwhelmed American engineers while the mobility of another object (the submarine) gave new opportunities to circumvent the material constraints of Greenlandic terrestrial ice. The Cold War was dominated by interest in onshore Greenland as an icy platform for US strategic operations, while for the Danes it emerged as a possible resource frontier for significant minerals such as uranium. The marine geographies of Greenland also captured interest.

From being framed as a possible strategic/geographical gap (the Greenland-Iceland-United-Kingdom (GIUK) gap) in the Cold War, the last decade has witnessed a growing appreciation of how to think of Greenland as possessing a stretchable quality with implications for the sovereign rights of Denmark/Greenland, its position as an Arctic nation, as well as Greenland's aspirations for nation-building and state formation (Nuttall 2014). Precisely when Greenland appears to be more confident in asserting greater economic and political autonomy, Denmark has moved to declare its position as a major Arctic state with a significant role in international diplomacy. Ocean depths and subsurface geological environments are important for the Kingdom of Denmark as an Arctic state whereas they assume quite different meanings for Greenlandic ambitions for independence. This stretchable quality of Greenland has implications in territorial and geopolitical ways – some for the consolidation of an Arctic state, some for the creation of a new Arctic state. Geologists and oceanographers have been integral to the mapping and making of offshore Greenland, the development of an emerging oil/gas sector, and an island with extended outer continental shelves, stretching all the way to the central Arctic Ocean. At the same time in Greenland, the government of Greenland's Ministry of Industry and Mineral Resources supports the gathering of geo-data through its department of geology to inform strategy-making, licensing, and the marketing of mineral resources for economic development.

Probing the Greenlandic seabed: Denmark as "Arctic state"

The Russian flag planting on the Arctic Ocean seabed in 2007 attracted attention because it employed a self-knowing colonial-imperial gesture of marking territory through the importation of a flag to a place far from human settlement. Dismissed by some as a publicity stunt and declared irrelevant under international law, it nonetheless unleashed a tsunami of commentary about "Arctic scrambles", suggesting that many instinctively understood the discursive-material significance of an event occurring far below the water, on the seabed and seen through the window of a submersible. The private provenance of the expedition added further intrigue to the expedition and possible Russian intentions towards the Arctic Ocean. Was Russia about to stake a claim to this oceanic space? Or was it a display of technological-exploratory chutzpah akin, as Russian commentators suggested, to the United States planting a flag on the surface of the moon in 1969? What the event demonstrated, though, was a different register of Arctic geopolitics to the one of Cold War Greenland. Rather than the ice sheet, the seabed becomes productive of the geopolitical. Exploring the depths of the Arctic continental shelf and submarine rights has emerged as a crucial element in the sovereignty politics of the Arctic Ocean. Legal regimes such as the UN Convention on the Law of the Sea (UNCLOS) have become conduits for bringing subterranean knowledge to the surface in order to justify and legitimate the roles of Arctic Ocean costal states.

In 2008, the five coastal states agreed to the Ilulissat Declaration, reiterating their collective desire to manage, in an orderly fashion, and within existing international legal frameworks, the issues and changes affecting the Arctic Ocean from climate change to shipping and fishing. The intervention was decisive in using international legal and geographical categories to establish, what in critical race studies is termed, a "somatic norm" – a naturalised domain for some people/bodies/ideas/states as opposed to others (Puwar 2001). The "somatic norm" at play revolves around the five Arctic Ocean coastal states stating that they are the rightful symbolic and geophysical occupiers of the maritime Arctic region. Using geographical proximity, discussions of Arctic geopolitics and governance privileges these states and their interests. Conversely "space invaders" in the form of other states and communities play a disturbing role in this space. Proximity, in this context, is working on two registers; geographical/geophysical and racial.

Geographically the Ilulissat Declaration privileged some Arctic states with Finland, Iceland and Sweden not being invited to the meeting. Racially, the Declaration was accompanied by a ceremony highlighting the role of white men as representatives of those five Arctic Ocean coastal states. No representatives from indigenous peoples' organisations were invited to participate let alone endorse the Declaration (a criticism levelled at the meeting by the Inuit Circumpolar Council, which later organised its own summit with a declaration on Inuit sovereignty). Although the Greenlandic premier Hans Enoksen was there, as were other Greenlandic politicians, they were participants as part of a state delegation rather than as representation of any assertions of indigeneity. The Ilulissat Declaration codes Arctic coastal states, and their white representatives, as a naturalised norm with the affect of making non-white bodies and non-coastal states (albeit with very different experiences and trajectories) unwelcome and alien, as well as ignoring and even erasing the presence of indigenous peoples.

The Danish government convened the meeting and selected the Greenlandic town of Ilulissat, a symbol of rapid climate change and tipping points. It did so against a backdrop of concerns about global climate change and a desire to promote its role as an Arctic Ocean coastal state, courtesy of Greenland. Six months later, Greenland hosted a referendum, which confirmed a popular desire for autonomy and further self-government including rights to administer and exploit the island's subsurface resources. Enacted in 2009 this ushered in a new era of Government of Greenland controlled mineral licensing in coastal lands and offshore licensed drilling zones (Nuttall 2012). As Greenlandic voters were casting their votes, representatives from Inuit and other indigenous communities met at the Inuit Leaders' Summit in Kuujjuaq, Nunavik, Canada to express their views about the Ilulissat Declaration in their own Circumpolar Inuit Declaration on Sovereignty in the Arctic:

> On 7 November, International Inuit Day, we expressed unity in our concerns over Arctic sovereignty deliberations, examined the options for addressing these concerns, and strongly committed to developing a formal declaration on Arctic sovereignty. We also noted that the 2008 Ilulissat Declaration on Arctic sovereignty by ministers representing the five coastal Arctic states did not go far enough in affirming the rights Inuit have gained through international law, land claims and self-government processes.
>
> (Inuit Circumpolar Council 2009)

Rather than "not go[ing] far enough" perhaps what they actually did do was to go deeper into the ocean and further offshore to cement their sovereign rights in the Arctic Ocean. While the Inuit Leaders' Summit manifested indigenous autonomy in the form of international legal recognition and land claims, the Arctic Five (A5) were codifying themselves as volumetric occupiers of the maritime Arctic region.

Arctic governance and geopolitics was further "tested" in the ministerial meeting of the Arctic Council in Nuuk in 2011 when the Arctic states and permanent participants considered the question of whether further states and organisations should be admitted as "observers" to the forum. Encouraged by Nordic member states, predominantly Asian countries such as China, Japan and South Korea had expressed interest in joining an overwhelmingly European group of existing observers. The ministerial meeting agreed to new guidelines, ensuring that all observers would have to recognise formally the sovereign rights of the eight Arctic states as well as the rights of their indigenous communities. Designed to reconcile the Arctic states A8 and A5 and Arctic communities (indigenous and non-indigenous), new rules of engagement with old and new observers were agreed upon. However, the admittance of new nation state observers in 2013, led to concerns from permanent participants that their collective voice and influence might be disrupted and even diluted by a reassertion of traditional state-centric governance.

The reassertion of state-centric governance in the Arctic has also manifested itself through scientific knowledge and practice about Arctic ice and seas. Governance, geopolitics and geophysics have co-constituted one another. The intersection of science and geopolitics in the Arctic remains significant in spaces and in areas such as the ocean depths and seabed. Science is a powerful mode of governance complicit with claims to authority and governance. The five coastal states have invested millions of dollars, roubles and kroner in the mapping of the continental shelves and seabed to demarcate the outer limits of their sovereign rights in the Arctic Ocean. These mapping projects become accomplices to traditional state-centric power, producing an Arctic geopolitics, which privileges the rights of the coastal state and the cartographic conventions of the modern nation-state. Denmark and its specialist agencies such as the Geological Survey of Denmark and Greenland in collaboration with international partners (including Sweden, Russia and Canada) remains active in mapping, charting and promoting this geo-vision of the Arctic Ocean as the rightful space of coastal states, as legitimated by the UNCLOS. The place of Greenland and its communities, however, remains ambivalent in this Arctic geopolitics. Made complicit with the ambitions of Denmark as coastal state and key Arctic player, most notably in declarations and articulations of Arctic political and cultural identity evident in the Kingdom of Denmark's Arctic strategy, in Danish submissions to the United Nations Commission on the Limits of the Continental Shelf (CLSC), and with increased funding for Danish Arctic research (Nuttall 2014), it is also a complex space with colonial encounters and Cold War histories and geographies, as we outlined earlier.

Without the continued constitutional relationship with Greenland, Denmark's identity as an Arctic Ocean coastal state would be jeopardised. Consequently, Denmark's "Arctic activism" (Rosamond 2015), combining a strong emphasis on Danish military presence allied with a public commitment to multilateralism and its special relationship with Greenland, needs to be seen in relation to the stakes at play for Danish sovereignty and identity. The Danish invitation to convene a meeting in Greenland also has local origins in Danish–Greenlandic politics. The then Danish foreign minister Per Stig Møller was deeply involved in climate-change diplomacy and had been active in launching a "Greenland dialogue" in 2005, which sought to draw attention at ministerial level to the implications of climate change for Arctic environments. In 2007, he warned an audience in London that climate change was geopolitically significant with implications for Arctic resource extraction, shipping, maritime policing and territorial ownership. Two months later, "evidence" for possible tension could be found in the Russian flag planting. The "scramble for the Arctic" discourse had begun in earnest, and the fate of the Greenlandic ice sheet has since been at the epicentre of popular and political discourse, as well as scientific narratives, about a warming world.

The Danish submission to CLCS was delivered in 2014 and the Minister of Foreign Affairs Martin Lidegaard noted at the time that:

The submission of our claim to the continental shelf north of Greenland is a historic and important milestone for the Kingdom of Denmark. The objective of this huge project is to define the outer limits of our continental shelf and thereby – ultimately – of the Kingdom of Denmark.

(Government of Denmark 2014)

Within the submission, the Danish government contended that the Lomonosov Ridge, which extends some 1100 miles across the Arctic Ocean and dividing into the Eurasia and Amerasia basins, is "both morphologically and geologically an integral part of the Northern Continental Margin of Greenland". The submission suggested that the outer continental shelf from the baselines of Greenland covers some 895,000 square kilometres. The Lomonosov Ridge is one area of the Arctic Ocean seabed that is of great interest not just to Denmark – the submission overlaps with the Canadian, Norwegian, Russian submissions to the CLCS. The extent of sovereign rights over the seabed is still to be determined but is likely to extend at least 350 nautical miles beyond the coastal baseline, and possibly further. The eventual settlement of outer continental shelf delimitation will involve multinational negotiation regardless of any recommendations from the CLCS. What this process might eventually reveal, however, is the outer limits of what the territorial extent of an independent Greenland may look like; an Arctic state stretching possibly to the North Pole itself, and an entity with its own sovereign rights over the exclusive economic zone and vast area of continental shelf.

Conclusion

For a time, Arctic geopolitics became rapidly reassembled around the politics of fear and even dread within a context of rapid climate change, sovereignty and territorial claims. Articles and books appeared with warnings about weak Arctic governance, resource and territorial scrambles, a "Cold Rush", and a "New Great Game" (e.g. Potapov and Sale 2009). The central Arctic Ocean – and what lies within and deep below it – was one such area. This subterranean territory invited a new era of colonial mapping, exploitation and administration. Informed by international law, attention turned to the provisions that allowed the coastal states to extend their sovereign rights to the outer continental shelves of their national territories. Defining themselves as an Arctic 5, Canada, Denmark/Greenland, Norway, Russia and United States reimagined themselves as coastal states with substantial interests and sovereign rights in the Arctic Ocean.

Since 2008, the Arctic Ocean coastal states have reinforced their special geographical relationship. While the seabed has been a very powerful material marker of that relationship, in more recent years the fate of the high seas of the central Arctic Ocean has provided further incentive. In 2014, Nuuk was host to a meeting on potential fishing activity in the central Arctic Ocean, which later led to an agreement by the coastal states to prohibit their vessels from fishing in the region until a regional fisheries agreement is in place including extra-territorial actors such as China, South Korea and the European Union.

The net result has been to reinforce, according to some Danish observers, a view of the Kingdom of Denmark as a "middle power" with a vested interest in the governance of the Arctic Ocean. Greenland's geographical qualities are clearly critical to this in terms of identification of Denmark/Greenland as a coastal state with specific sovereign rights. What complicates this understanding is the growing autonomy of the government of Greenland (self-government) and its formal competence to take ownership and control over its subsurface resources both on and offshore. Granted home rule in 1979, the introduction of self-government in June 2009 has been followed by increasing desires for greater autonomy from Denmark in which discussion of foreign and security policy often come to the fore. The 2011 Strategy for the Arctic reinforced the role of the Danish military in terms of protecting Danish sovereignty in Greenland and the wider Arctic region.

Greenlandic politics has been closely influenced by the role and extent of the extractive sector and whether the government of Greenland should be working more closely with foreign companies and investors to help generate revenue streams, to help fund a shift away from economic reliance on Denmark. Resource stakeholders (politicians, government bodies and institutions, local businesses, multinational companies) are imagining and making the resource frontier in Greenland, and the extractive industries are re-imaging onshore and offshore areas as being of great potential, as part of a wider trans-national "New North" closely connected to the world economy (Nuttall 2012; 2013). Despite a dip in global commodity prices, as well as other global processes, the subterranean and the ocean depths nonetheless remain critical for Greenlandic notions of nation-building and state formation. Plans for mining and oil development projects, even with their accompanying social and environmental impact assessments, as well as discussions of the environmental and social impacts of seismic surveys and mineralogical mappings, involve extractivist discourses and spatial technologies of power that privilege particular techno-centric and economic views of the Greenlandic environment and do not take into account local community perspectives on human–environment relations (Nuttall 2015).

Particular places become emptied of human presence and activity and are reimagined as resource spaces marked out for economic development and accompanying ambitions for Greenlandic state formation. Within the discursive space created by the idea and formulation of a resource frontier as a "zone of unmapping", to use Anna Tsing's phrasing (2005), a diverse range of actors have become engaged in the production, mediation and reproduction of different kinds of Greenlandic futures, something which involves a new mapping and classification of Greenlandic spaces filled with possibility, opportunity and ambition. While it has become a stated aim of recent Greenlandic governments to "extract" revenue from hydrocarbon projects, mining activities and energy and industrial development, official plans have provoked highly charged political and social debates within Greenland about the nature and desirability of such a development and how it may redefine the nature of place and territory. At the same time, contested perceptions and understandings of the environment have become increasingly apparent with concerns expressed by local people and grassroots organisations, as well as international environmental and conservation groups, about threats to community viability, to wildlife and to biodiversity. Local understandings of human–environment relations are ignored, especially within social and environmental impact assessments for possible projects, and local experience and knowledge, as well as local histories of past mining activities, are erased by the production of technical knowledge and in political and industry discourses about Greenlandic environments and subsurface resources.

All of this has social and political implications within Greenland, as well as for relationships within the Kingdom of Denmark, and for Greenland's place in the world. As we have argued in this chapter, to locate Greenland within a critical geopolitics involves a consideration of the science and politics of and about ice, land and water, as well as the subsurface and Greenland's depths and widths: this is vital for contemporary understanding of how the subsurface is imagined, probed, mapped and politicised, how territory is thought about, and what happens at the intersection of both Greenlandic political discourse and extractive industry narratives surrounding resource development and its possibilities, the emergent public responses to it and the growth of social movements and assemblage of local protest, debates over decision-making processes and the extent and nature of public participation, and the growing influence of corporate transnationalism over Greenlandic politics and even everyday life (Nuttall 2015). Our concern with how Greenland is not just placed but how it is materialised within a critical geopolitics, however, also illustrates a broader process of the reimagining of the Arctic as a resource frontier and a space for economic possibility, and the way in which ice sheets, mountains, waterways, ocean depths and subterranean geologies are enrolled in geopolitical imaginaries and narrative concerning resource futures, and the hopes and ambitions, as well as the anxieties and resistances to which this gives rise.

Notes

1 In 1961, SIPRE was merged with the Arctic Construction and Frost Effects Laboratory to create a Cold Regions Research and Engineering Laboratory based in Hanover, NH.
2 There was a substantial European–US partnership in Greenland ice core research between the 1950s and 1980s. For further details see Elzinga, 2011.

References

Bowman, I. (1949). Geographical interpretation. *Geographical Review* 39(1), pp. 355–370.
Bruun, J. and Medby, I. (2014). Theorising the thaw: Geopolitics in a changing Arctic. *Geography Compass* 8(12), pp. 915–929.
Bridge, G. (2013). Territory, now in 3D!. *Political Geography* 34, pp. 55–57.
Clark, N. (2013). Geopolitics at the threshold. *Political Geography* 37, pp. 37–40.
Cruikshank, J. (2004). *Do Glaciers Listen? Local Knowledge, Colonial Encounters, and Social Imagination*. Vancouver: University of British Columbia Press.
Dodds, K. (2010). Flag planting and finger pointing: The law of the sea, the Arctic and the political geographies of the outer continental shelf. *Political Geography* 29, pp. 63–73.

Dodds, K. and Nuttall, M. (2016). *The Scramble for the Poles.* Cambridge: Polity.

Dunbar, M. (1950). Greenland during and since the Second World War. *International Journal* 5(2), pp. 121–140.

Elden, S. (2013). Secure the volume: Vertical geopolitics and the depth of power. *Political Geography* 34, pp. 35–51.

Elzinga, A. (2011). Some aspects of in the history of ice core drilling and science from IGY to EPICA. In: C. Ludecke, L. Tipton-Everett and L. Lay, eds, *National and Trans-National Agendas in Antarctic Research from the 1950s and Beyond.* Columbus, OH: Byrd Polar Research Center.

Government of Denmark. (2014). Denmark and Greenland will today file a submission regarding the continental shelf north of Greenland. Available at: http://a76.dk/cgi-bin/nyheder-m-m.cgi?id=1418619842%7Ccgifunction=form [Accessed 17 February 2017].

Grosz, E. (2008). *Chaos, Territory and Art.* New York: Columbia University Press.

Hough, P. (2013). *International Politics of the Arctic.* London: Routledge.

Inuit Circumpolar Council (2009): A Circumpolar Inuit Declaration on Sovereignty in the Arctic 2009. Available at: http://inuit.org/icc-greenland/icc-declarations/sovereignty-declaration-2009/. [Accessed 21 May 2017].

Kinney, D. (2013) Selling Greenland: The big picture television series and the army's bid for relevance during the early Cold War. *Centaurus* 55(3), pp. 344–357.

Martin-Nielsen, J. (2012). The other cold war: The United States and Greenland's ice sheet environment, 1948–1966. *Journal of Historical Geography* 38(1), pp. 69–80.

Martin-Nielsen, J. (2013). *Eismitte in the Scientific Imagination: Knowledge and Politics at the Centre of Greenland.* New York: Palgrave.

Nuttall, M. (2010). *Pipeline Dreams: People, Environment, and the Arctic Energy Frontier.* Copenhagen: IWGIA.

Nuttall, M. (2012). Imagining and governing the Greenlandic resource frontier. *The Polar Journal* 2(1), pp. 113–124.

Nuttall, M. (2013). Zero-tolerance, uranium and Greenland's mining future. *The Polar Journal* 3(2), pp. 363–383.

Nuttall, M. (2014). Territory, security and sovereignty: The Kingdom of Denmark's Arctic strategy. In: R. Murray and A. Nuttall, eds, *International Relations and the Arctic: Understanding Policy and Governance.* New York: Cambria Press.

Nuttall, M. (2015). Subsurface politics: Greenlandic discourses on extractive industries. In: L. Jensen, and G. Hønneland, eds, *Handbook of the Politics of the Arctic.* Cheltenham: Edward Elgar, pp. 105–127.

Orlove, B., Wiegandt, E., and Luckman, B., eds. (2008). *Darkening Peaks: Glacier Retreat, Science, and Society.* Berkeley, CA: University of California Press.

Petersen, N. (1998). Negotiating the 1951 Greenland Defense Agreement: Theoretical and empirical aspects. *Scandinavian Political Studies* 21(1), pp. 1–28.

Potapov, E. and Sale, R. (2009). *The Scramble for the Arctic.* New York: Frances Lincoln.

Puwar, N. (2001). The racialised somatic norm and the senior civil service. *Sociology* 35(3), pp. 651–670.

Robinson, M. (2006). *The Coldest Crucible: Arctic Exploration and American Culture.* Chicago, IL: University of Chicago Press.

Rosamond, A. (2015). The Kingdom of Denmark and the Arctic. In: L. Jensen and G. Hønneland, eds., *Handbook of the Politics of the Arctic.* Cheltenham: Edward Elgar, pp. 501–516.

Roucek, J. (1951). The geopolitics of Greenland. *Journal of Geography* 50(6), pp. 239–246.

Roucek, J. (1983). The geopolitics of the Arctic. *The American Journal of Economics and Sociology* 42(4), pp. 463–471.

Said, E. (1993). *Culture and Imperialism*. London: Vintage.

Squire, R. (2016). Rock, water, air and fire: Foregrounding the elements in the Gibraltar–Spain dispute. *Environment and Planning D: Society and Space* 34(3), pp. 545–63.

Strandsbjerg, J. (2012). Cartopolitics, Geopolitics and boundaries in the Arctic. *Geopolitics* 17(4), pp. 818–842.

Tsing, A. (2005). *An Ethnography of Global Connection*. Princeton, NJ: Princeton University Press.

Thompson, C. (2005). *Making Parents: The Ontological Choreography of Reproductive Technologies*. Boston, MA: MIT Press.

Conclusion

The opportunities and challenges of Greenlandic paradiplomacy

Kristian Søby Kristensen and
Jon Rahbek-Clemmensen

In December 2016, Vittus Qujaukitsoq, Greenland's minister for foreign affairs, gave a controversial interview to a Danish newspaper in which he reacted to a string of controversial issues about the American presence in Greenland that had surfaced over the past years (including the lack of compensation for environmental consequences and general fiscal remuneration). Highlighting that he reacted to "75 years of accumulated frustration and impotence", Qujaukitsoq declared that Denmark did not treat Greenland fairly, that information provided by US authorities could not be trusted, and that Greenland should be given the authority to pursue its own interests vis-à-vis the United States and other foreign entities (Hannestad 2016a). If Copenhagen and Washington did not accede, Greenland should gain independence and expel the US. In a speech a few days before, Qujaukitsoq advocated closer bilateral cooperation between Greenland and Canada and the US without Danish interference (Hannestad 2016b; Qujaukitsoq 2016).

The incident, which caused uproar in Danish and Greenlandic media, illustrates some of the main conclusions of the present volume: that Greenland can use its postcolonial legacy and high politics importance to increase its paradiplomatic room for manoeuvre, but that postcolonialism and high politics also entail specific challenges that other paradiplomatic actors rarely face (*Berlingske* 2016; Hansen 2016a; 2016b; Kristiansen 2016). Furthermore, domestic and international politics are intertwined in Greenland as domestic politics and national identity are shaped by international affairs, while they conversely shape Greenland's foreign policy. Greenland's paradiplomacy affects Denmark's bilateral relations with other nations and the general politics of the Arctic.

Greenland's connection to Denmark is looser than that of most paradiplomatic actors and Nuuk consequently has significant room for manoeuvre, even though the island's foreign and security policy formally remains within Copenhagen's purview. It is often difficult to determine whether an issue is domestic or foreign policy, which means that Denmark is never absent and Danish politics spill over into Greenlandic debates, as exemplified by Kevin Foley's analysis of the China debate.

The postcolonial debt and the possibility of Greenlandic independence clearly give Nuuk some leverage vis-à-vis Copenhagen and external actors. For instance, as Mikkel Runge Olesen argues, Nuuk used the postcolonial past to influence the 2003 Thule Air Base agreement, even though the negotiations were originally a bilateral US–Danish issue (see also Kristensen 2005). However, postcoloniality concurrently hampers Greenland's manoeuvrability in international affairs, although these disadvantages seem small compared to the advantages. Maintaining a position as champions of the Inuit cause entails an active diplomatic effort to find a *modus operandi* with other Inuit actors with conflicting goals. As Hannes Gerhardt shows, Greenland's vision differs from that espoused by other Inuit actors, most importantly the ICC, as it emphasises the importance of a state-centric, Westphalian order in the Arctic. Furthermore, as Camilla Sørensen argues, Chinese diplomats do not feel comfortable dealing with a self-governing entity with a clearly stated desire for independence.

Similarly, the presence of high politics interests also strengthens and inhibits Greenland's room for paradiplomatic manoeuvring. The great powers have an interest in Greenland, which thus gives Nuuk an opportunity to forge new relations and to attract investments. Greenland's geostrategically important location between North America and Asia has made the island an important security interest for the United States especially after the Second World War and the rising importance of the Arctic and the economic opportunities that follow attract other foreign actors, most importantly China, to Greenland. This, in turn, will not make Greenland less important to Washington. Greenland simultaneously plays an increasingly important role in Danish foreign and security policy as it helps strengthen the ties between Copenhagen and Washington, which forces Denmark to ensure that relations with Nuuk remain amenable. However, the importance of Greenlandic issues also restricts Nuuk's room for manoeuvre. For example, as both Camilla Sørensen, Kevin Foley, and Mikkel Runge Olesen highlight, the American security interests in Greenland make it difficult for China to invest in Greenland, as Danish authorities are afraid to give China leverage that could jeopardise the American position in Greenland and complicate US–Danish relations. Furthermore, as Jon Rahbek-Clemmensen shows, the rising importance of the High North has attracted Copenhagen's attention and Denmark now pursues an Arctic foreign policy that places less emphasis on bilateral relations with Nuuk and where Greenland is given less leash to pursue its own foreign policy.

The chapters thus show that the world is interested in Greenland and that Greenland has significant paradiplomatic room to shape its own, independent foreign policy. Nuuk plays an active role in Arctic politics, both vis-à-vis other Arctic and non-Arctic nations and in forums, such as the Arctic Council, but also with other sub-state groups, most importantly the Inuit. Arctic politics is more than just the Ilulissat Declaration and the interaction between the eight Arctic states and their foreign partners. Sub-state and non-state actors continue to play a role in the ordering of the region and Greenland will perhaps be the most important sub-state actor in the years to come, because of its unique position as a semi-autonomous entity with some independent foreign policy clout. For instance, Inuuteq Holm Olsen and Jessica Shadian show that Greenlandic paradiplomacy, namely its insistence on representing itself at Arctic Council meetings, helps expose tensions within the Westphalian structure of the Council. Greenlandic policymakers have already demanded reforms of the Arctic Council that gives more influence to sub-state actors and it seems likely that Inuit groups will make similar moves in the years to come (Sørensen 2016).

Finally, Greenland not only has an effect on world politics, world politics also has an effect on Greenland. The Greenlanders became aware of their own identity in their meeting with the world and relations to foreign powers continue to shape how politics are discussed in Nuuk. In their analysis of the uranium debate, Kristian Søby Kristensen and Jon Rahbek-Clemmensen showed that Greenlandic debates about international issues are simultaneously debates about different visions for an independent Greenland. Jens Heinrich tracked this process historically and showed the notion of an independent Greenland was a result of the meeting with the outside world. As Greenland's independence project has matured and its political interaction with external actors has widened and, as Ulrik Pram Gad and Marc Jacobsen demonstrate, it has come to mirror itself vis-à-vis a plethora of different actors. The relationship to the United States has become part of Greenland's postcolonial legacy and Iceland's past experiences serve as an ideal, to name just two examples. More broadly speaking, Klaus Dodds and Mark Nuttall argue that deeply embedded ideas and narratives about Greenland – its geophysical and climatic characteristics and the opportunities and dangers it contains – also shape how the world approaches the island and how it too thinks about itself. Global politics is thus a crucial part of Greenland's independence project and Nuuk will continue to work to empower itself to play an important role in the politics of the globe's northernmost region.

References

Berlingske. (2016). Danmark og Grønland driver hver sin vej. (18 December 2016), p. 40.

Hannestad, A. (2016a). For os er det 75 års opsparet frustration. *Politiken* (16 December 2016), p. 4.

Hannestad, A. (2016b). Grønland roser Trumps politik i Arktis. *Politiken* (14 December 2016), p. 9.

Hansen, N. (2016a). Atassut: Uklogt træk af Vittus Qujaukitsoq. *Sermitsiaq*. Available from: http://sermitsiaq.ag/atassut-uklogt-traek-vittus-qujaukitsoq. [Accessed 11 Jan 2017].

Hansen, N. (2016b). Randi tager afstand fra Qujaukitsoqs udtalelser. *Sermitsiaq*. Available from: http://sermitsiaq.ag/randi-tager-afstand-qujaukitsoqs-udtalelser. [Accessed 11 Jan 2017].

Kristensen, K. (2005). Negotiating base rights for missile defence: The case of Thule Air Base in Greenland. In: S. Rynning, K. Kristensen and B. Heurlin, eds, *Missile Defence: International, Regional and National Implications*. Abingdon: Routledge, pp. 183–207.

Kristiansen, C. (2016). Løkke: Vi har et godt forhold til Grønland. *Politiken* (14 December 2016), p. 9.

Qujaukitsoq, V. (2016). Greenland, Canada, and the United States: the Arctic Potential. Speech given at Arctic Circle Quebec. Available at: http://naalakkersuisut.gl/~/media/Nanoq/Files/Attached%20Files/Udenrigsdirektoratet/Arctic%20Circle/Tale_VQ_Arctic_Circle_Quebec_ENG.pdf. [Accessed 1 January 2017].

Sørensen, H. (2016). Olsvig ønsker magtopgør i Arktisk Råd. *KNR*. Available from: http://knr.gl/da/nyheder/olsvig-%C3%B8nsker-magtopg%C3%B8r-i-arktisk-r%C3%A5d. [Accessed 6 Mar 2016].

Index